REVISE FOR

EDEXCEL

MODULAR

science

Graham Booth

Bob McDuell

John Parker

Higher

Heinemann Educational Publishers
Halley Court, Jordan Hill, Oxford, OX2 8EJ
a division of Reed Educational & Professional Publishing Ltd

Heinemann is a registered trademark of Reed Educational & Professional
Publishing Ltd.

OXFORD   MELBOURNE   AUCKLAND   JOHANNESBURG
BLANTYRE   GABORONE   IBADAN   PORTSMOUTH (NH)   CHICAGO

First published 2002

ISBN 0 435 57891X

06 05 04 03 02
10 9 8 7 6 5 4 3 2 1

Edited by Sara Hulse

Typeset and illustrated by Tech Set Ltd

Printed and bound in Spain by Edelvives

Index by Paul Nash

Acknowledgements
Cover photo by Digitalvision
Photos by PhotoDisk

The publishers have made every effort to trace the copyright holders, but if
they have inadvertently overlooked any, they will be pleased to make the
necessary arrangements at the first opportunity.

Contents

Introduction

We have written this book to help you prepare in the best way possible for your exams.

Throughout the two years of your course you will sit a total of 12 module tests and two terminal exam papers. You can find tips on preparing for these in the *Examiners Tips* section (pages 149–152).

Make sure you find out the dates for your tests and exams to give your self plenty of time to get ready for them.

Work through the chapter for the module you are about to sit. Pace yourself – do one spread at a time and look through your notes made in class.

Try the questions at the end of every spread to check you really understand the topic.

Check your answers under *Answers to end of spread questions*. Go back over anything you found difficult.

Try the practice module questions at the end of each module. these are similar to your module test questions. Try to do them in the time allowed.

Check your answers in the Answers to exam style questions at the end of the book.

When you come to prepare for the terminal exams, look through all your notes and try the Exam style questions in this book.

Most importantly, Good luck!

The Authors

The human body – action and control

1

This module explains some of the most important life processes that take place in the human body, including digestion, control, excretion and defence.

The first topic in the module explains how the digestive system breaks down food to produce soluble nutrients suitable for absorption into the bloodstream.

A control system allows body activities to run smoothly. This is known as the nervous system and is covered in the second topic.

Topic three outlines the excretory functions of the kidneys and the final topic describes the structure and functions of the blood and the skin and explains their transport and defence functions.

The digestive system

The process of digestion

A meal is full of useful nutrients but the food needs to be broken down into small soluble molecules that can pass into the bloodstream. The process of breaking food down happens in stages and is called **digestion**.

Food is broken down by digestive juices, which contain the substances responsible for this breakdown – known as **enzymes**.

The action of enzymes • • •

Enzymes are sometimes called **biological catalysts**. An enzyme is a substance that:

■ speeds up a chemical reaction;

■ has an **active site** that allows a chemical (substrate) of a suitable shape to 'fit in';

■ results in new substances being formed;

■ is unchanged at the end of a reaction, ready to be used again;

■ works best at an **optimum temperature** (37°);

■ works best at an **optimum pH**.

Each enzyme is a protein molecule that has a special shape. A substrate of the correct shape can fit into the **active site** and be broken down into small soluble molecules. Enzymes are highly specific.

Above this temperature reactions are slower. If the temperature is too high, enzymes are destroyed or **denatured**. Below 37°C reactions also slow down and if the temperature is too low the enzymes remain **inactive**.

The special functions of each digestive enzyme are shown in the table below:

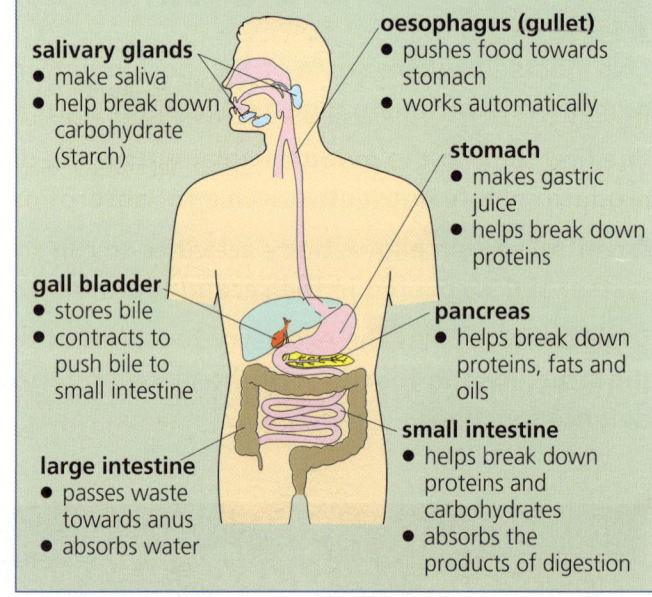

The human digestive system

salivary glands
● make saliva
● help break down carbohydrate (starch)

oesophagus (gullet)
● pushes food towards stomach
● works automatically

stomach
● makes gastric juice
● helps break down proteins

gall bladder
● stores bile
● contracts to push bile to small intestine

pancreas
● helps break down proteins, fats and oils

large intestine
● passes waste towards anus
● absorbs water

small intestine
● helps break down proteins and carbohydrates
● absorbs the products of digestion

Remember that you will be examined on the function of the parts of the digestive system (not the names or their position!).

Enzyme	Produced by...	Acts on...	Reaction takes place in...	To produce...
carbohydrase	salivary gland	starch	mouth	sugar
protease	gastric gland	protein	stomach	chains of amino acids
carbohydrase	pancreas	starch	small intestine	sugar
lipase	pancreas	fats and oils	small intestine	fatty acid and glycerol

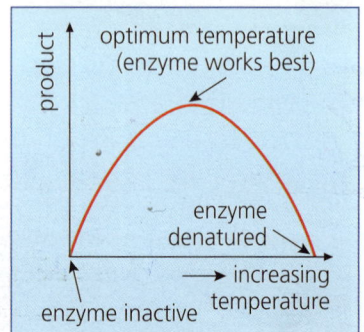

The effect of temperature on an enzyme-catalysed reaction

The function of bile in digestion • • • • • •

Bile helps digestion because:

■ It breaks up fats and oil into tiny pieces, giving them a large surface area. This allows lipase to change them into fatty acids and glycerol more efficiently. This process is **emulsification**.

■ Bile is an alkali. It neutralises the acid from the stomach as it passes into the small intestine, allowing the enzymes in the small intestine to work at a suitable pH.

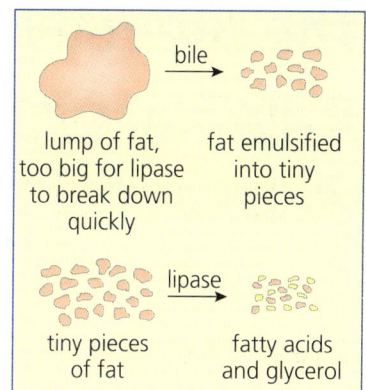

Absorption of nutrients • • • • • • • • • •

When the food reaches the small intestine, it has been broken down into molecules small enough to be absorbed into the bloodstream.

Special adaptations of the small intestine help in the efficient absorption of nutrients:

■ The small intestine is very long, and has large numbers of villi, giving it a large surface area through which nutrients can be absorbed.

■ Each villus has the special function of absorption. Villi have a pointed shape and are very thin. The lining of the villus is only one cell thick, nutrients can easily pass into the blood capillaries.

> Each of the villi is known as a villus. There are millions of them, so huge amounts of nutrients can be absorbed.

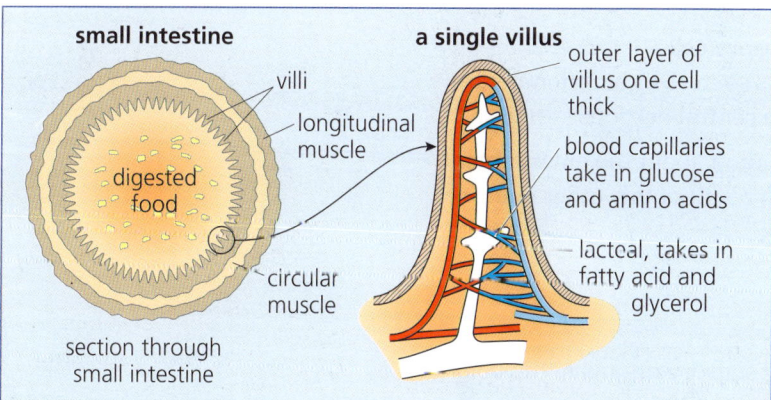

Nutrients pass through the wall of each villus into the blood. The nutrient molecules move from where they are in high concentration to where they are in low concentration. This is called **diffusion**.

> What happens in the large intestine?
> • water absorbed into bloodstream
> • makes the undigested food waste more solid

Questions

1 What conditions allow an enzyme to work at its fastest rate?

2 What happens to an enzyme at a very high temperature?

3 What happens to an enzyme at a very low temperature?

4 What happens to protein in the mouth? Give a reason for your answer.

5 Describe two features of the small intestine that give a high surface area for the absorption of nutrients.

The nervous system

You need to know ●

✔ the structure and function of a nerve cell;

✔ the functions of key parts of the eye;

✔ some effects of a range of drugs on the nervous system.

The nerve cells ● ● ● ● ●

Each nerve cell is known as a **neurone**. Throughout the nervous system neurones communicate by **electrical impulses**.

Some important facts:

- Fat on the outside of a neurone insulates it so that the electrical impulse remains in the cell.

- Electrical impulses pass along the axon.

- When an impulse reaches a muscle, the muscle contracts.

- A receptor is found only on a sensory cell.
  A receptor responds to a stimulus such as light or temperature and sends an impulse back to the central nervous system.

- Sensory neurones supply information about the environment inside and outside the body. As a result of this other neurones respond.

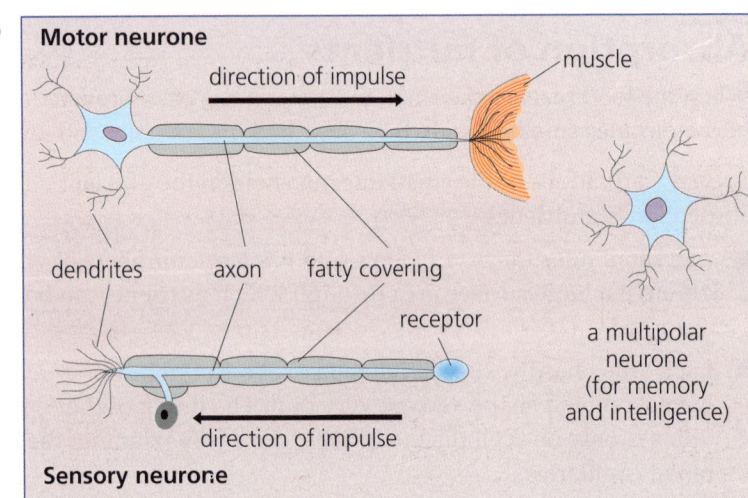

The eye

The eye is one of the most important sensory organs in the body.

The iris reflex

The amount of light that enters the eye has to be controlled. This is done by the **iris reflex**. Light reaches the retina at the back of the eye. As a result, the iris muscles cause the pupil to change diameter. This happens automatically, and very fast. The brain is not involved in this action so it is known as a **reflex action**. A wide pupil (**dilated**) allows more light to enter the eye. A narrow pupil (**constricted**) reduces the amount of light entering the eye.

sensory neurone (in retina) → impulses sent through optic nerve → central nervous system → relay neurone → motor neurone → impulses sent → iris muscles respond

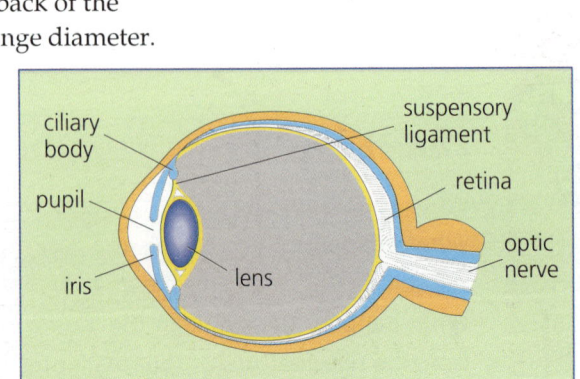

Structure of the eye

What is the function of the retina? • • • •

When we look at something, we focus its image on the **retina**, a layer at the back of the eye. Some neurones (**rods**) of the retina detect black and white, and some (**cones**) detect colour.

Electrical impulses travel from the retina through the optic nerve to the brain. When these impulses are received we can see.

> If the focus is not on the retina, then the image is blurred. Lenses can be used to correct this.

The effects of drugs on the nervous system •

Many substances have an effect on the body. Some drugs can be used to help relieve symptoms when people are ill. Some drugs can be misused and harm human health.

Drug	Problem
paracetamol – a painkiller that works by preventing the impulses of sensory neurones reaching the brain, so no pain is experienced	overdose damages the liver, and can cause death
heroin – a painkiller that can be given in a refined form known as morphine for serious conditions, especially terminal illness	very addictive
caffeine – found in coffee, this speeds up body reactions and so is known as a stimulant – it makes people more active	speeds up the heart rate can result in heart problems
barbiturates – slow down reactions and are known as sedatives – make people less active – used as sleeping tablets	can be addictive
alcohol – a sedative, slowing down reactions	reaction times are slower, hence the dangers of drink-driving; destroys brain cells with regular use and can damage the liver by a condition known as cirrhosis
solvents – affect the working of the body in a number of ways, all bad	destroy nerve cells (neurones) and destroy the lungs, e.g. alveoli are badly damaged
tobacco – has an effect on the nerve cells but no useful function	addictive, increases risk of cancer, causes respiratory disease, e.g. bronchitis, and heart disease

Drugs can change behaviour and personality. Many crimes are drug-related and violence is a regular outcome. As a result of drug misuse, people have a lower resistance to viral infection.

Questions

1. Describe the role of (a) phagocytes and (b) lymphocytes in defending the body.
2. How is a red blood cell adapted for the transport of oxygen?
3. What are the conditions of the lens, suspensory ligaments and ciliary body as the eye focuses a near object?
4. Why is it important that light entering the eye is controlled?
5. What type of drug is alcohol? Give one problem that can result if someone drinks too much alcohol over a number of years.

The kidneys

The kidneys – the body's filter system • •

The kidneys are part of the urinary system:

- blood, at high pressure, reaches **Bowman's capsule**;
- this forces some **urea**, **glucose** and **water** molecules out of the **glomerulus** into the **first convoluted tubule** – this process is **ultrafiltration**;
- proteins and blood cells do not pass into the tubule, because they are too big to pass through the capillaries of the glomerulus;
- blood capillaries also run close to the first convoluted tubule;
- 100% of glucose and some water are reabsorbed here;
- **reabsorption** is the movement of substances from the kidney tubule back into the blood;
- urea, a toxic substance, continues to the collecting tubule, then the ureter and finally the bladder, where the urine leaves the body via the **urethra** at infrequent intervals.

All glucose is reabsorbed as it passes through the first convoluted tubule. This is by a process called active transport.

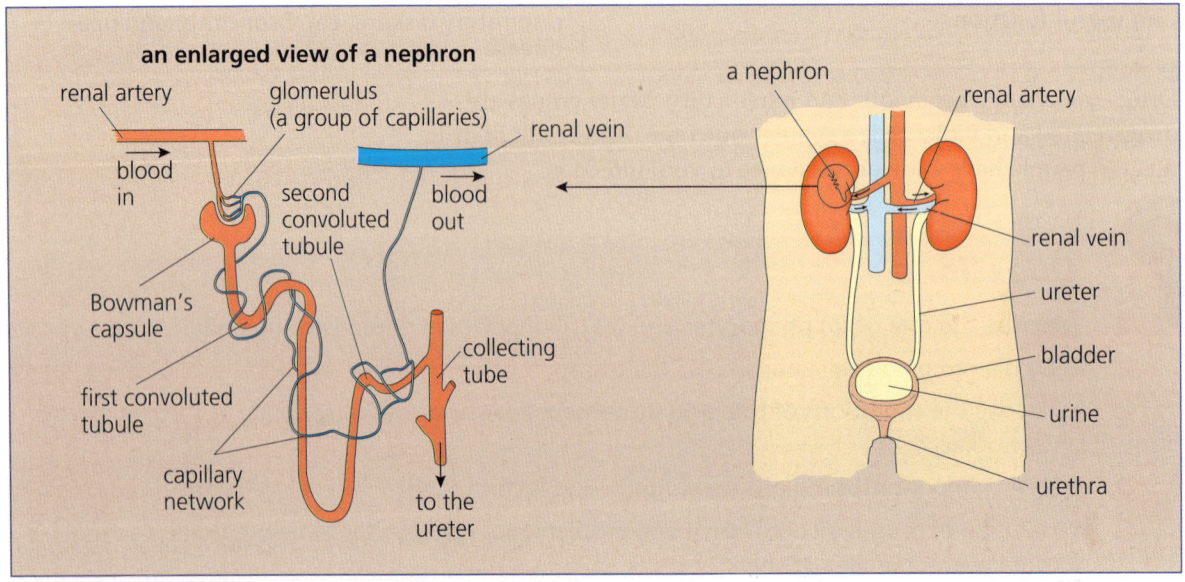

The human urinary system

Why we need to excrete urea

Amino acids in very high concentration in the blood are dangerous. The liver converts excess amino acids into a less toxic substance, urea. Even urea is toxic, so must not be allowed to build up in the blood. The kidneys filter it out of the blood, and it is excreted in the urine.

> Learn the terms ureter and urethra carefully. Many exam candidates mix them up – make sure that you don't!

Controlling the water content of the blood

The hormone ADH (antidiuretic hormone) controls the water content of the blood. ADH:

- ■ is produced by the pituitary gland;
- ■ reaches the kidney via the blood;
- ■ acts on the kidney tubule so that more water is reabsorbed into the blood.

ADH secretion varies in different temperatures:

> Students often mix up the kidneys and the liver. Remember that the liver makes the urea and the kidneys excrete it.

In warm conditions			
more water lost as sweat	more ADH produced	more water reabsorbed back into blood	less water lost in urine
In cold conditions			
less water lost as sweat	less ADH produced	less water reabsorbed back into blood	more water lost in urine

Homeostasis

This is the maintenance of a constant internal environment:

- ■ Too much of a chemical, e.g. water, in the body causes an imbalance. Not enough of a chemical also results in imbalance. It needs to be regulated, so ADH aids homeostasis.
- ■ Body temperatures above or below 37°C would cause severe problems with all enzyme-catalysed reactions. Homeostasis makes sure that optimum is maintained. Release more energy in your cells and your temperature increases. Release less energy in the cells and temperature decreases.

> Use the table to learn about the effects of ADH. Take care with the use of more and less when describing ADH and water – you need to understand everything in the table to answer successfully.

Questions

1 What is reabsorption?

2 Explain why the fluid in the tubule increases in concentration as it passes through the first covoluted tubule.

3 Why is homeostasis necessary?

4 List the parts in the kidney that a urea molecule passes through before leaving the body in urine.

Defence and protection

You need to know •

✔ the structure of the skin;

✔ how the skin helps to control body temperature by sweating, as well as by vasoconstriction and vasodilation;

✔ how the skin helps to defend against infection;

✔ how the skin acts as a sense organ.

✔ the components of blood and their functions;

✔ the role of blood in transport around the body;

✔ about the clotting of blood.

Your skin is very important in preventing microorganisms entering your body. Your blood plays an important role in repairing skin when it becomes damaged and destroying any microorganisms that do get into your body.

The structure of the skin • • • • • • • • • • •

The skin consists of a number of different tissues, each having a different function.

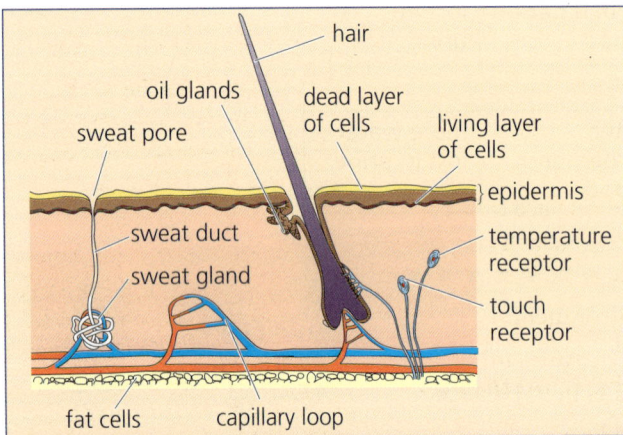

The structure of the skin

An outside layer of dead cells forms a tough barrier to prevent microorganisms entering the body. Oil glands produce sebum, which both lubricates the hairs and kills some microorganisms using the enzyme lysozyme.

Touch receptors and nerve receptors are situated at the ends of sensory nerve cells. These receptors, embedded in the skin, send impulses to the brain, so that we can feel things that we touch, and can feel how hot they are.

The components of blood • • • • • • • • • • •

The different components of blood include plasma, red blood cells, white blood cells and platelets (see diagram).

Plasma	White blood cells	Red blood cells
the liquid part of the blood	destroy microorganisms that invade the body	do not have a nucleus
carries the blood cells	are found as two different types	have a biconcave shape (like a doughnut)
carries all hormones		contain haemoglobin
carries nutrients such as glucose		transport oxygen
carries waste substances such as urea		

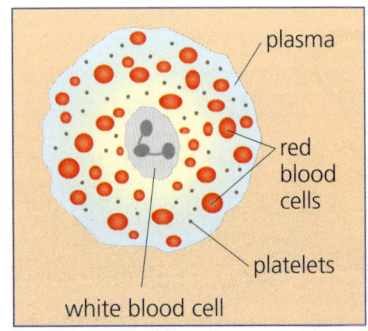

This is what blood looks like under a microscope

Blood clotting

Under normal body conditions blood flows along the vessels. When a vessel is damaged, the blood seems to change from a liquid to a solid, forming a clot.

A blood clot is needed for two reasons:

- to prevent the loss of blood from a damaged vessel;
- to prevent the entry of microorganisms into the bloodstream.

The platelets produce fibres made of fibrin. These fibres form a mesh and trap the red blood cells, like fish in a net. This helps to form a scab.

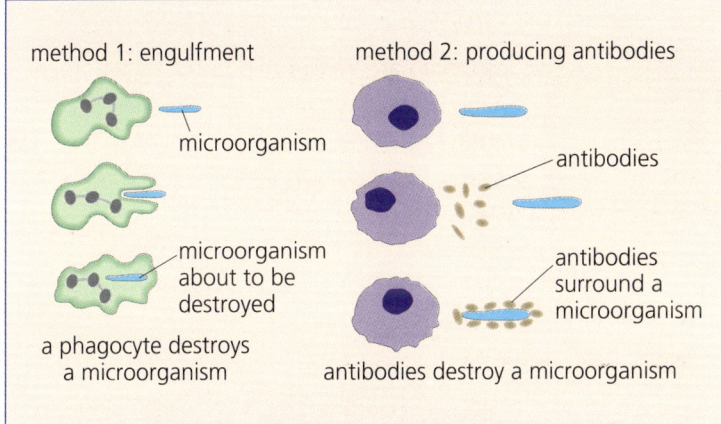

Red blood cells do not have a nucleus, and so have a "dip" in the middle. This gives them a high surface area, so more oxygen can be transported. There are millions of red blood cells in the human body, allowing a lot of oxygen to be transported.

Haemoglobin is the substance that helps oxygen molecules cling to the red blood cells

How do white blood cells destroy microorganisms?

Without white blood cells harmful microorganisms would be able to attack us more successfully. We would have less of a defence.

Control of body temperature

Sweating

Sweat glands produce sweat, which flows onto the surface of the skin and evaporates. Evaporation needs heat energy, which is removed from the skin and so the body cools down.

Vasoconstriction and vasodilation

- Capillaries in the skin can be empty or full of blood. At the start of each **capillary loop** there is a ring of muscle (**arteriole**).

- When an arteriole is opened (dilated) blood flows into the capillaries, the skin feels warmer, so more heat can be lost from the skin by convection and radiation. This is **vasodilation**.

- When an arteriole is closed (constricted) blood flows into the core of the body, so less heat can be lost from the skin by convection and radiation. This is **vasoconstriction**.

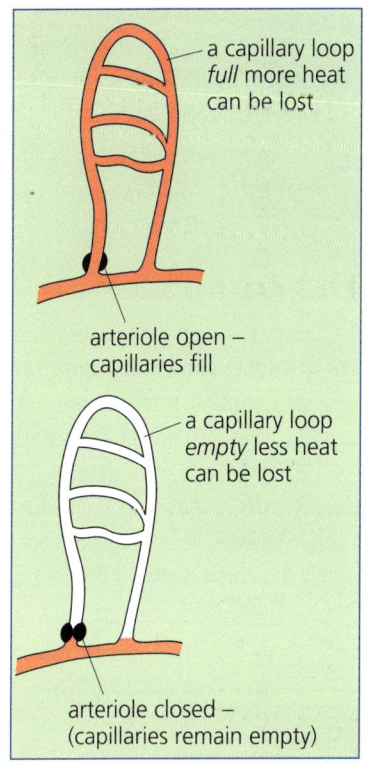

a capillary loop *full* more heat can be lost

arteriole open – capillaries fill

a capillary loop *empty* less heat can be lost

arteriole closed – (capillaries remain empty)

Questions

1. What is the function of lysozyme?
2. How does sweat cool us down?
3. During which process do arterioles dilate?

Practice module test

You will have 17 minutes to answer these questions

1 Which of the following structures provide a high surface area for the absorption of nutrients into the blood?

A villi

B salivary glands

C gall bladder

D pancreas

2 Saliva contains the enzyme:

A protease

B carbohydrase

C lipase

D maltase

3 Which of the following statements is not true?

A plasma helps the blood to clot

B the kidneys produce urea

C the large intestine absorbs water

D phagocytes engulf bacteria

4 Which of the following drugs slow down nervous impulses?

A stimulants

B painkillers

C hallucinogens

D sedatives

5 The following statements describe the process of sweating but they are in the wrong order:

1 water in sweat evaporates

2 the body cools down

3 the sweat glands produce sweat

4 sweat lies on the surface of the skin

Which of the following gives the correct order?

A 2 1 4 3

B 4 1 3 2

C 3 4 1 2

D 3 4 2 1

6 Which of the following is **not** true of a red blood cell?

A it contains haemoglobin

B it is circular, so that it can pass through blood capillaries easily

C it has a nucleus

D it does not have a nucleus

7 Hormones are transported to the target organ:

A in the blood plasma

B by red blood cells

C by white blood cells

D by platelets

8 Paracetamol is a painkiller. It can be dangerous because:

A it is addictive

B it can harm the liver if taken in overdose

C it reduces reaction time

D it produces hallucinations

9 The enzyme lipase breaks down fats into:

A amino acids

B fatty acid and glycerol

C amino acids and glycerol

D sugars

10 Which of the following parts of the eye work together to focus the image on the retina?

A the cornea and the pupil

B the pupil and the lens

C the cornea and the lens

D the conjunctiva and the lens

11 The lens in the eye changes shape when we look at an object at different distances from the eye. Which row of the table below is correct when the eye is focused on an object close by?

	Lens	Suspensory	Ciliary body
A	long and thin	tight	contracted
B	long and thin	loose	relaxed
C	round	loose	contracted
D	round	tight	contracted

12 The figure shows fluid passing through the first convoluted tubule in a kidney. Which statement is a correct explanation of what happens between A and B?

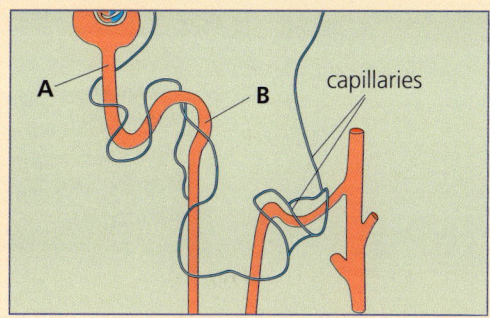

  A water enters the tubule from the blood capillaries
  B glucose is being added to the tubule
  C protein is being reabsorbed
  D glucose is being reabsorbed

13 Which of the following three statements about bile are correct?
  1. Bile is alkaline and neutralises acid from the stomach
  2. Bile increases the surface area of fats
  3. Bile contains the enzyme lipase

  A 1 and 2
  B 2 and 3
  C 1 and 3
  D just 3

14 Which of the following statements correctly describes vasodilation?
  A arterioles open to allow blood to the core of the body
  B arterioles close to allow blood to the core of the body
  C arterioles open to allow blood close to the skin surface
  D arterioles close to allow blood close to the skin surface

15 A person enters a brightly lit room. Which of the statements below correctly describes what happens?
  A light reaches the iris – the iris changes the pupil size
  B light reaches the retina – the retina sends impulses to the iris – the iris dilates
  C light reaches the retina – the retina sends impulses to the optic nerve – motor neurones cause the iris to dilate
  D light reaches the retina – the retina sends impulses to the optic nerve – sensory neurones cause the iris to dilate

16 The flow diagram shows what happens when damage to a blood vessel stimulates the platelets to begin clotting. Which of the following sequences is the correct order for the process of blood clotting?

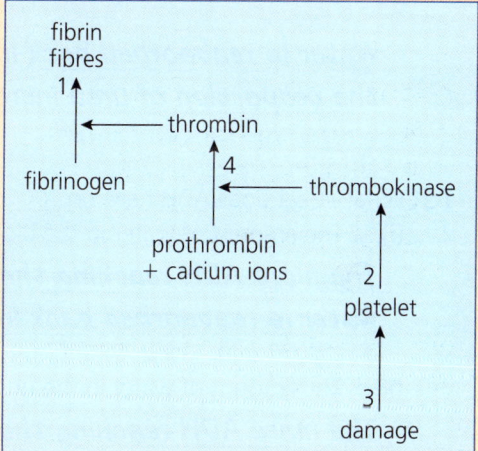

  A 1 2 3 4
  B 2 3 4 1
  C 4 2 3 1
  D 3 2 4 1

Answers to these questions can be found on pages 143–147

Getting it right

1 The diagram shows a kidney nephron.

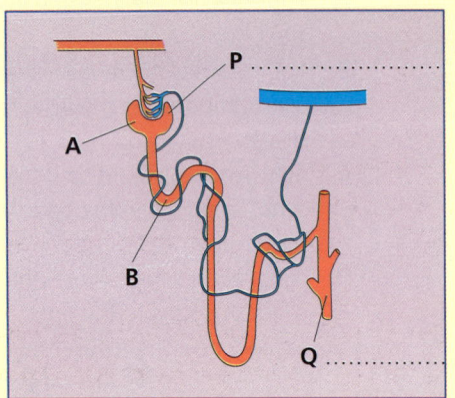

(a) Name parts P and Q.

P = Bowman's capsule. Q = collecting duct. **[2]**

(b) Explain why the concentration of the glomerular filtrate increases from A to B.

Water is reabsorbed back into the blood so the proportion of urea increases. **[2]**

(c) What effect does ADH have on:
(i) The amount of water in the blood?

The more ADH reaching the kidney, the more water is reabsorbed back into the blood. **[3]**

(ii) The amount of water in the urine?

The more ADH reaching the kidney, the less water in the urine, because more water goes back into the blood. **[3]**

(d) Which structure takes fluid from
(i) the kidney to the bladder?

Ureter **[1]**

(ii) the bladder to the toilet?

Urethra **[1]**

Inheritance and survival

2

Genes control the characteristics of living things and have a major influence on survival. This module is about genes, the mechanism of inheritance and the effect of the environment on the expression of genes.

The first topic in the module is about human chromosomes and how characteristics are passed on through sexual reproduction. It includes an outline of the roles of meiosis and mitosis in reproduction and development.

The role of DNA in variation and inheritance is the subject of the second topic. This leads to topic three, which includes an outline of genetic engineering, the major technological advance of our time, with full consideration of the ethics involved. Species have changed over millions of years. The fourth topic explores aspects of natural selection and survival that have led to the evolution of different species. Humans can harm their own existence and that of other living things by polluting the environment. The effects of pollution are explored in this final topic

Chromosomes and variation

You need to know ●

✔ the numbers of chromosomes in human body cells and sex cells;
✔ the outcome and significance of mitosis and meiosis;
✔ that sex is controlled by specialised chromosomes – the X and Y chromosomes;

✔ the role of human sex cells in fertilisation;
✔ how sexual reproduction results in variation across the human population;
✔ the functions of male and female sex hormones.

Human body cells ● ● ● ● ● ● ● ● ● ● ● ● ● ● ●

A human adult consists of around 5 billion cells. Different cells do different jobs but there is one common structure in every cell – the **nucleus**. The human nucleus contains 23 pairs of **chromosomes**, known as the **diploid number**. Each chromosome is like a row of **genes**, each of which control a particular human characteristic.

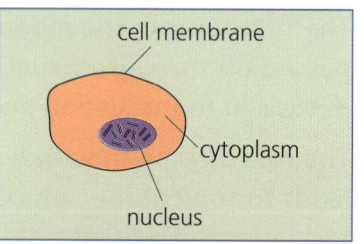
cell membrane
cytoplasm
nucleus

Human sex cells ● ● ● ● ● ●

Sex cells are produced by both men and women and carry the chromosomes responsible for male and female characteristics. There are two different sex chromosomes in the nucleus – X and Y. A male has an X and a Y chromosome. A femalehas two X chromosomes. Each sex cell carries 23 single chromosomes 22 + X or 22 + Y). This single set of 23 chromosomes is called the **haploid number**.

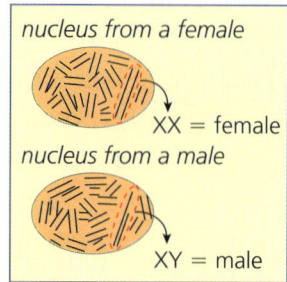
nucleus from a female
XX = female
nucleus from a male
XY = male

An X chromosome is longer than a Y chromosome. This means there are more genes on one side (X) than on the other (Y).

Cell division ● ● ● ● ● ● ● ● ● ●

Some cells in the human body can divide. There are two forms of cell division, each having a different function.

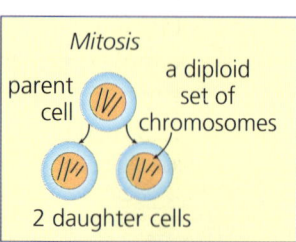
Mitosis
parent cell
a diploid set of chromosomes
2 daughter cells

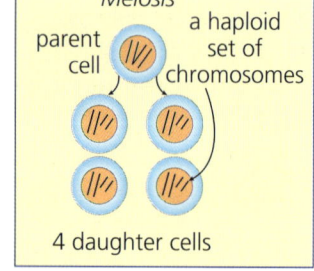
Meiosis
parent cell
a haploid set of chromosomes
4 daughter cells

Mitosis

■ A cell divides to form **two daughter cells**.
■ Each chromosome replicates itself so that the complete set of chromosomes is copied.
■ Each cell is identical – a **clone**.

■ Each cell is **diploid**, because it has **23 pairs of chromosomes**.
■ This type of cell division is used for growth and repair, e.g. replacement of skin cells.

Meiosis

■ A cell divides twice to form **four daughter cells**.
■ Each chromosome replicates itself so that the complete set of chromosomes is copied.
■ Each pair of chromosomes exchanges **alleles** from one chromosome of the pair to the other, resulting in chromosomes at the end of cell division being different.

■ Each cell is **haploid**, because it has **23 single chromosomes**.
■ This type of cell division is used to produce **gametes** (sex cells) – so every gamete produced by a male or female is always different, so meiosis is a source of variation.

Fertilisation

This is the fusion of the male sex cell (**sperm**) with the female egg cell (**ovum**) to form one cell, a fertilised egg cell or **zygote**. Males produce sperm that carry *either* 22 single chromosomes plus X *or* 22 single chromosomes plus Y. Females can only produce eggs that carry 22 single chromosomes plus X. The sex of a baby depends only on whether the sperm that reaches the egg first carries X or Y.

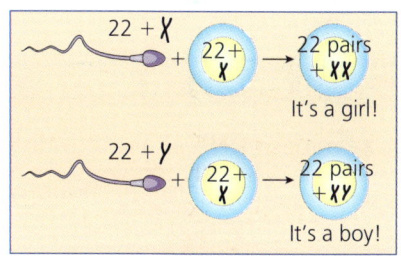

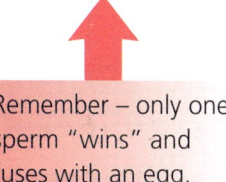

Remember – only one sperm "wins" and fuses with an egg.

Sexual reproduction and variation

A man may produce 200 million sperm in one ejaculation. Each sperm produced is different to the next. Every egg produced by a woman is different. This is why children look noticeably different to their parents – this is known as **variation**.

Both parents pass on genes to a baby. A gene may control a feature such as eye colour. We all know that there are different eye colours (blue, brown, grey, etc.). Each colour is a different expression of the same gene, known as an **allele**. For each feature of a baby, one allele is inherited from the mother and one from the father. Each allele can be either **dominant** or **recessive**.

If a dominant allele is inherited, the characteristic always shows up in the baby. If an allele is recessive, the characteristic is only expressed if the baby inherits two recessive alleles. When one parent passes on a dominant allele and the other a recessive allele, then the dominant version of the gene shows up in the baby.

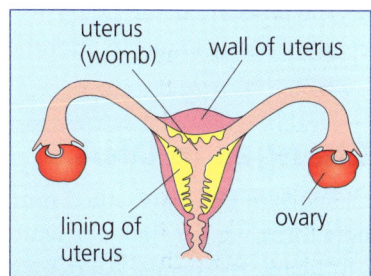

The sex hormones

Hormones are produced by endocrine glands. Each hormone is like a chemical message that is transported through the body in the blood. The sex hormones are very important in giving a person the characteristic features of their sex – for example the deep voice of a teenage boy or the enlargement of breasts in a teenage girl.

Name of hormone	Where is the hormone produced	How does it reach its target	What effect does it have?
testosterone	testes	in blood	• deep voice • sperm produced • wide shoulders • pubic hair • facial hair
oestrogen	ovaries	in blood	• produced eggs • hips become wider • breasts become bigger • pubic hair • lining of uterus builds up during menstrual cycle
progesterone	ovaries	in blood	keeps the lining of the uterus in place during pregnancy

These hormones cause the **secondary sexual characteristics** to develop – the changes that take place during the teenage years.

Questions

1 Which chromosomes in humans control the sex of a person?

2 How many chromosomes are found in a human (a) body cell, (b) gamete?

3 Which type of cell division results in variation in offspring?

4 Complete the sentences by writing in a number.
Mitosis produces _____ daughter cells.
Meiosis produces _____ daughter cells.

Genes and variation

You need to know •

✔ about DNA as a unit of inheritance;

✔ the mechanism of monohybrid inheritance;

✔ how to predict ratios of genotypes and phenotypes in offspring;

✔ that the environment can affect inherited characteristics;

✔ that some diseases can be inherited;

✔ some implications of the Human Genome Project;

✔ the causes and consequences of changes in DNA (mutation);

✔ that DNA can be artificially changed by gene modification.

DNA – the secret of life! • • • • • • • • • • • •

Genes control human characteristics and are made from **D**eoxyribo**N**ucleic **A**cid – DNA. DNA consists of:

■ two strands, twisted to form a double helix;

■ strands that are linked together like the rungs of a ladder;

■ the links are made by **bases** that join together in specific pairs – **thymine** (T) links to **adenine** (A) and **cytosine** (C) links to **guanine** (G);

Changes in DNA • • • • • • • • • • • • • • • • •

Changes in DNA are called **mutations**. A gene may change into a new form, e.g. an allele for white petals may change to one producing red petals. An extra chromosome may even appear, or extra genes may be added or deleted.

Changes in DNA can be caused by:

■ radiation (X-rays and gamma rays);

■ ultraviolet light;

■ substances produced when tobacco is burned.

the bases bond together at the middle

T links to A
G links to C

only the pairings shown can link

Monohybrid inheritance • • • • • • • • • • • •

Two organisms may have only one difference, e.g. eye colour.

Allele for brown eyes = B (dominant)

Allele for blue eyes = b (recessive)

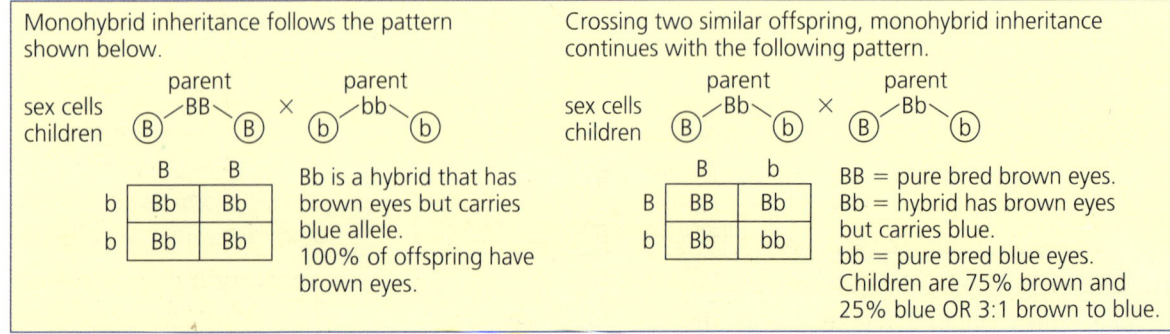

Other important terms used in genetics

Homozygous – an individual carries two alleles that are the same. In humans, one allele is from the mother and one is from the father.

Heterozygous – an individual carries two different alleles. In humans, one allele is from the mother and one is from the father.

Genotype – all of the alleles in an individual, i.e. all dominant alleles and all recessives, whether they are expressed or not. Harry's genotype is BB, Sinita's genotype is Bb.

Phenotype – only the alleles that are expressed in the person. Sinita's phenotype is B.

Genetic diseases

Some genes can be faulty and cause a **genetic disease**, which can be inherited. The chances of inheriting a genetic disease are affected by whether the faulty allele is dominant or recessive, and by which chromosome the gene is located on.

How does the environment affect genes?

Genes control the manufacture of proteins in the cells. Proteins are used to make structures. For example, the weight of a baby at birth depends on a number of factors:

- the genes of the developing fetus;
- the diet of its mother during her pregnancy;
- other substances that pass into the blood of the fetus during development, including drugs such as nicotine;
- the health of the mother.

Asexual reproduction

Some organisms can reproduce without using any sex cells. Parent cells divide to produce cells that are genetically identical to the parent. The new cells are **clones** of the parent. Examples include:

- Spider plants (Chlorophytum) – drooping stems hang towards the soil and develop roots. Mini-plants develop that root into the soil, forming clones of the parent plant.
- During their life cycle, female greenfly produce young without any fertilisation. Their cells simply divide to form the new generation.

Both the spider plant and the greenfly reproduce asexually, but sexual reproduction is also used, which helps to produce offspring more quickly.

Type 1 – faulty allele dominant
N = faulty allele (dominant) **n** = normal allele (recessive)
individuals **NN** and **Nn** would have genetic disease, **nn** would be normal
many individuals in the population would have the disease.
Huntington's Chorea is passed by a **dominant** faulty allele. symptoms – muscles move automatically, not under voluntary control and there is mental deterioration.

Type 2 – faulty allele recessive
N = normal allele (dominant) **n** = disease allele (recessive)
NN is normal. **Nn** is a carrier. Only **nn** individuals have the genetic disease.
few individuals in the population would have the disease.
cystic fibrosis is passed on by a **recessive** faulty allele. symptoms – thick mucus is secreted in the lungs, pancreas and intestines, which causes pain, digestive problems, fluid build-up in the lungs.

a spider plant *reproducing asexually*

female greenfly

many young produced *identical* to *parent*

Questions

1 In a DNA molecule, which base links to (a) cytosine, (b) thymine?

2 What was the aim of the Human Genome Project?

3 Define asexual reproduction.

Selection, survival and evolution

You need to know

✔ the ways in which organisms are adapted to their environment;

✔ factors that affect numbers in the populations of a variety of species;

✔ the principles of natural selection leading to evolution;

✔ the principles of selective breeding (artificial selection);

✔ about pollution and human survival.

How can adaptations help an organism to survive?

Organisms that successfully live in their habitat have **adaptations** that make them suited to their environment. The examples below illustrate typical adaptations.

The organisms in your exam questions may be different but they will display similar adaptations.

Arctic fox	Fennec fox
Habitat – cold	Habitat – warm
• small ears to reduce surface area for heat loss • long fur, traps air, good insulation • white fur, good camouflage, not seen by predators or prey	• large ears to increase surface area for heat loss • short fur, not good insulation, easy to lose heat • brown fur, good camouflage, not seen by predators or prey

Predators and prey

Predators are **carnivores** – animals that eat other animals. This could lead to problems – if a predatory species kills all the prey in an area then it would cause its own death. The key part of the predator–prey relationship is that numbers remain in balance. An example is the relationship between the lynx (predator) and the snowshoe hare (herbivorous prey). The following graph shows the numbers of each population from 1910 to 1970.

Take care when answering questions about animals that eat plants like cacti. They are not carnivores – they are herbivores!

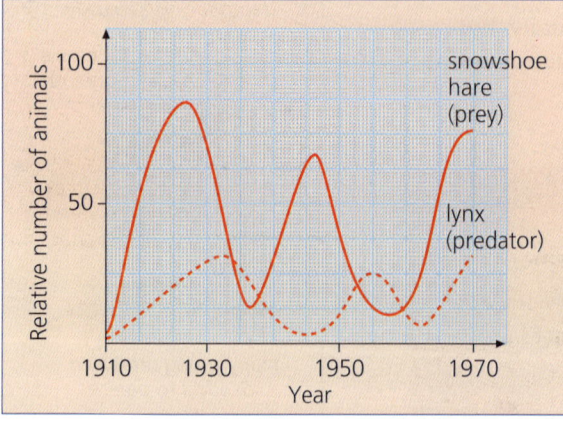

- each spring, plants grow so there is more food for the hares;
- the hare population breeds and numbers increase;
- predators have more hares to kill, so hare numbers decrease;
- the lynx (predator) population has a lot of food available and reproduces, so increasing in numbers;
- more lynx need more food, but there are now fewer hares, so their population decreases again;

Competition

This happens when organisms have the same needs. With a limited amount of food available, there is not enough for all organisms. Those better adapted to 'fight for their share' are good competitors. Others may perish.

Natural selection

Natural selection was first proposed by Charles Darwin. He observed the differences across species throughout the world. The key part to understanding this theory is that within a species there is **genetic variation**.

- different alleles may give individuals a better adaptation to the environmental conditions;
- some individuals die;
- some individuals survive and go on to pass on their alleles to the next generation, which now have the genetic advantage.

Why do some members of a species have different adaptations?

1. Sexual reproduction produces different combinations of alleles.
2. New genes and alleles appear occasionally – these are **mutations**.

Mutations appear spontaneously. A mutation can be good or bad for the species. Some bad mutations resulted in the death of a species. This is **extinction**. Good mutations resulted in changes to species over the years. This is **evolution** – it continues to take place today.

Is there evidence that evolution has taken place?

Yes! Organisms died millions of years ago. Most rotted down completely but some did not. The remains can be found today and are known as **fossils**. Many fossils are so well preserved that we can find out accurate details of their structure.

Selective breeding

Selective breeding means that people choose organisms with good features to cross. Offspring are tested and only the best ones are used for further breeding.

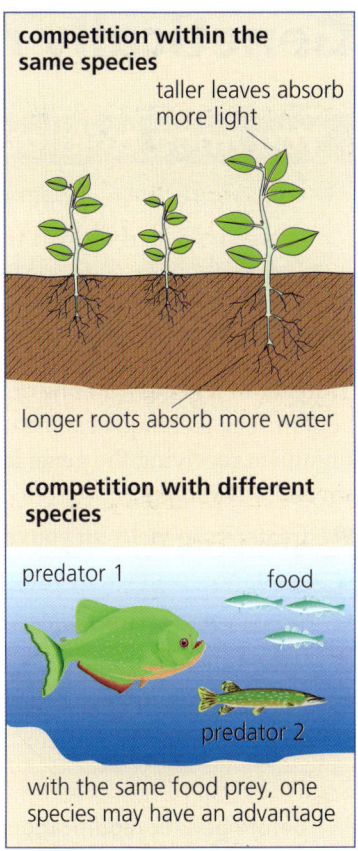

competition within the same species

taller leaves absorb more light

longer roots absorb more water

competition with different species

predator 1

food

predator 2

with the same food prey, one species may have an advantage

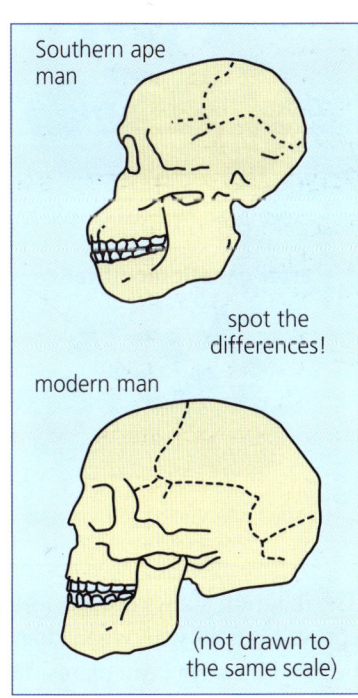

Southern ape man

spot the differences!

modern man

(not drawn to the same scale)

Evolution has changed the shape of the human skull.

Questions

1. How is a cactus adapted to survival in a desert?
2. Suggest three reasons why herbivores may die out.
3. What is natural selection?
4. Give two ways in which plants of the same species compete.
5. What is a mutation?

Genetically modified organisms

You need to know •

✔ that genetically modified organisms are produced by two transfer of gene(s) from a donor to recipient;

✔ The benefits of genetic modifications

Transfer of a gene from one organism to another can be of great benefit. The technique can be used to give **disease resistance** to the organism receiving the gene. In the case of crop plants, there can be several advantages:

- greater crop yield for conventional crops;
- resistance to attack by fungi;
- resistance to attack by insects;
- better rate of photosynthesis;
- crop products that lack allergy-causing chemicals, e.g. cereals may be made gluten-free.

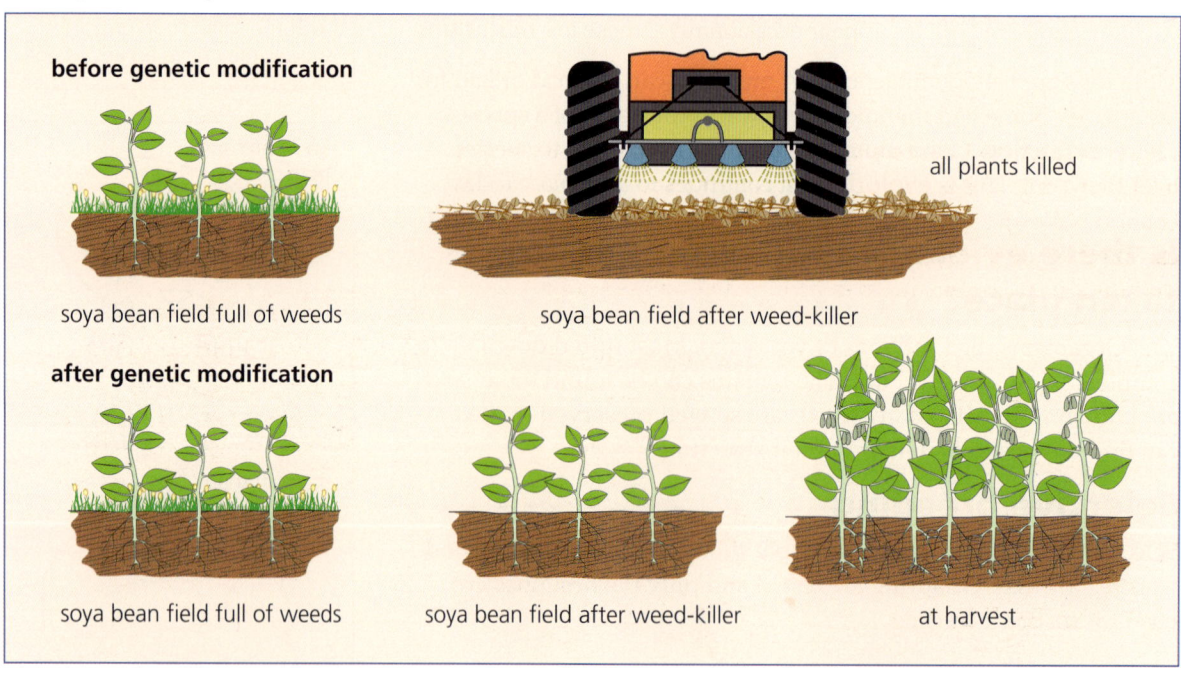

before genetic modification

soya bean field full of weeds

soya bean field after weed-killer

all plants killed

after genetic modification

soya bean field full of weeds

soya bean field after weed-killer

at harvest

The diagram shows how, before genetic modification, the farmer sprays his field with weed killer – this not only kills the weeds but also the soya bean plants. This means that the farmer has to use mechanical methods such as hoeing instead, which is more expensive. After the soya bean plants have been genetically modified, they are resistant to the weed killer, giving high crop yields with no weeds.

Think about the ethics of gene transfer. Is it right to change the genes of a species? We need to consider the advantages and disadvantages.

Pollution and human survival

You need to know ●

✔ that human activities cause pollution; ✔ the cause and effects of acid rain.

A bigger human population has increased needs. More food, more buildings, more cars, more products from industry are required. The list seems endless.

These requirements result in substances and processes that are harmful to the environment – this is pollution. Humans are responsible!

You may be given data to analyse. Look out for population increases and data involving additional needs, resulting in pollution.

Acid rain ●

Acid rain is produced when:

■ fossil fuels such as coal and petrol are burned;

■ sulphur dioxide is released into the air;

■ the sulphur dioxide combines with water vapour in the air;

■ this forms sulphurous acid (acid rain).

Acid rain can corrode buildings, kill plants and destroy aquatic life – even complete food chains can be destroyed.

Larger human populations mean increased traffic – this can also lead to the formation of acid rain. Exhaust gases from vehicles contain other gases as well as sulphur dioxide – carbon monoxide, carbon dioxide and nitrogen oxides – these can also form acid rain. Pollution from car exhaust can be reduced by using a catalytic converter – this reduces the toxic emissions from car exhausts. Alternatively, we could use vehicles that do not give off toxic substances, e.g. electric or solar-powered vehicles.

Other pollutants – like smoke – can also cause problems. Dust particles in smoke can cover plants and reduce photosynthesis.

lots of carbon monoxide and nitrogen oxides

This car needs a catalytic converter

Questions

1 What is pollution?

2 What causes acid rain?

3 What is the main reason for increases in pollution?

Practice module test

You will have 17 minutes to answer these questions

1   Given below is a sequence of bases in a DNA molecule. What are the missing bases?

A A T G C C
T T A _ _ _
A = adenine, T = thymine, G = guanine, C = cytosine

    A   T T A
    B   G G C
    C   C G G
    D   C C G

2   Evolution:

    A   takes place when a species becomes extinct
    B   no longer takes place
    C   consists of changes to species over millions of years
    D   does not rely on mutations to take place

3   Differences in the weight of people is caused by:

    A   just the environment
    B   just the genes
    C   genes + plus the environment
    D   just food

4   Choose the correct line from the following table to show how fertilisation produces a baby boy.

	sperm	+	egg	=	fertilised egg
A	XX	+	XX	=	XXXX
B	X	+	Y	=	XY
C	X	+	X	=	XX
D	XY	+	Y	=	XXY

5   In a flower the allele for red petals (R) is dominant to the one for white petals (r). If plants of the genotypes rr and Rr are crossed, what would the next generation be?

    A   50% red, 50% white
    B   50% pink, 50% white
    C   75% red, 25% white
    D   25% red, 75% white

6   Ladybirds are predators of greenfly. The food of greenfly is the sap of plants. Ladybirds increase in numbers:

    A   before an increase in greenfly
    B   after an increase in greenfly
    C   when plants die down
    D   during the winter

7   Which of these can cause mutations in skin cells?

    A   sweating too much
    B   eating too much protein
    C   too much muscular activity
    D   ultraviolet light

8   A sperm cell carries the following number of chromosomes:

    A   22
    B   23
    C   24
    D   46

9   Acid rain is dangerous because:

    A   it makes the pH of lakes very high
    B   by destroying plants it destroys food chains
    C   it causes skin cancer
    D   it increases the mineral level in soil

10   During one year a lake became more acidic as a result of high rainfall. Which graph below (**A**, **B**, **C** or **D**) shows this?

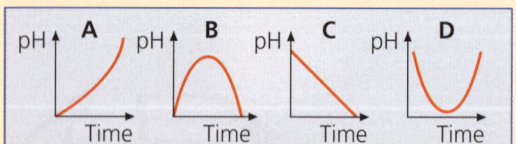

11   A black female guinea pig was given to a breeder. B = black (dominant), b = white (recessive). The breeder needed to find out if the new guinea pig had a genotype of BB or Bb. He bred the new guinea pig with a white female. Which line shows the correct results if the new guinea pig carried the white allele?

    A   50% black, 50% white
    B   100% black
    C   100% white
    D   75% black, 25% white

12 The diagram shows a cell dividing. Which of the statements below is true of this cell division?

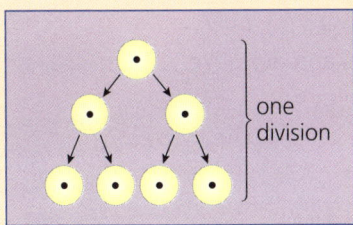

one division

- **A** cells are being cloned
- **B** the cell is dividing by meiosis
- **C** the cell is dividing by mitosis
- **D** the new cells are diploid

13 Two different variants of a species of snail live in an area. One has red bands and one has green bands. Both snails live in grassland. Which of the statements below is most likely to come true?

- **A** the red ringed snails have better camouflage and so will increase in number
- **B** the green ringed snails have better camouflage and so will increase in number
- **C** predators will not be able to see the red snail easily so more will survive
- **D** predators will be able to see the green snails easily so more will be killed

14 Which of the following statements about the type of cell division, meiosis, is true?

- **A** the cells produced are identical to the parent cells
- **B** the cells produced are all haploid
- **C** the cells produced are all diploid
- **D** the cells produced have identical chromosomes to the parent cell

15 Which one of the following is correct?

- **A** genotype + environment = phenotype
- **B** phenotype + environment = genotype
- **C** genotype + phenotype = environment
- **D** genotype = only the alleles that are expressed

16 Tall pea plants are crossed with short pea plants. The cross is shown by the genetic diagram. T = tall (dominant), t = short (recessive). The next generation are:

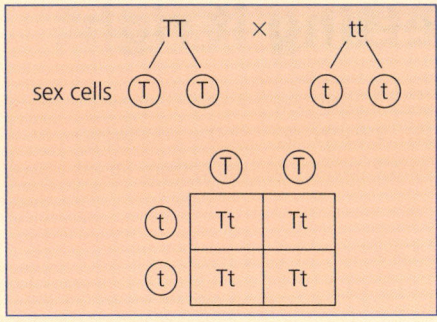

- **A** all medium height
- **B** 100% short
- **C** 100% homozygous
- **D** 100% heterozygous

17 Which of the following sequences shows part of a DNA molecule?

- **A** A A T G C G
  T T C C T C
- **B** C C G C T A
  T T A T A C
- **C** C G C A A T
  G C G T T A
- **D** C G C A A T
  C G C A A T

18 The graph below shows the results of one form of cell division. One cell divides and results in the production of a number of identical cells. Which of the following statements explains the results shown on the graph?

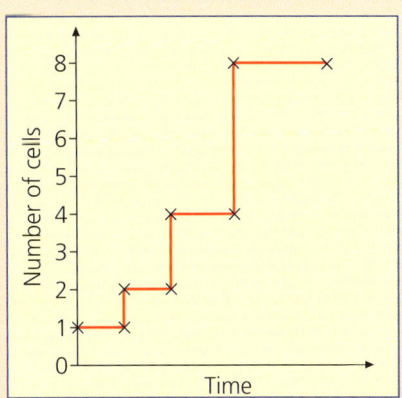

- **A** beginning with one cell, two meiotic divisions are shown
- **B** beginning with one cell, two mitotic divisions are shown
- **C** beginning with one cell, there were three mitotic divisions
- **D** beginning with one cell, there were three meiotic divisions

Answers to these questions can be found on pages 143–147

Getting it right

1 Cystic fibrosis is a genetic disease passed on by a recessive allele. The symptoms include the production of thick mucus in the lungs and digestive tract, which causes serious health problems.

N = allele for normal mucus production

n = cystic fibrosis trait

The diagram below shows the family tree of the Smith family. Use information from the family tree to help you answer the questions.

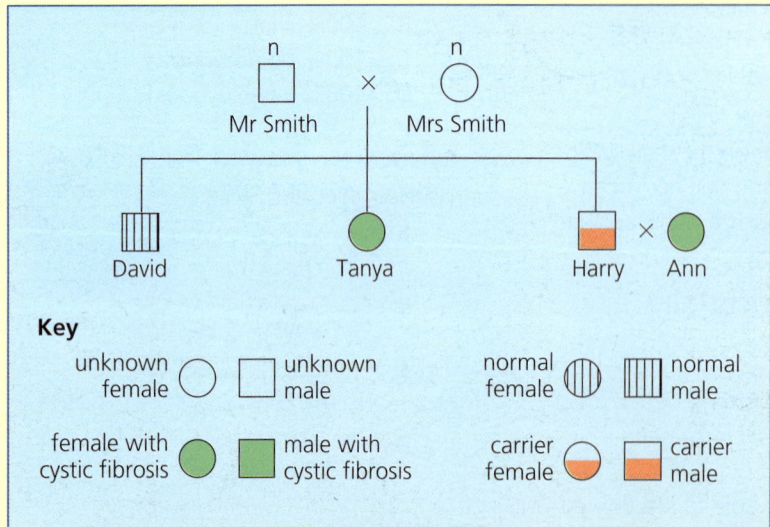

Key

unknown female ○	unknown male □	normal female ⊕	normal male ⊞	
female with cystic fibrosis ●	male with cystic fibrosis ■	carrier female ◒	carrier male ◩	

(a) What are the genotypes of Mr and Mrs Smith? Explain how you worked out each genotype.

> *Tanya is a female with cystic fibrosis, so both Mr and Mrs Smith passed on "n". Harry is a carrier so one parent passes on "N" and one "n". At least one parent has "N". David produces normal mucus, is not a carrier so has "N" from both parents.*
> *Mr Smith's genotype is Nn. Mrs. Smith's genotype is Nn.*   [4]

(b) What are the chances of Harry and Ann producing a girl with cystic fibrosis? Show your working.

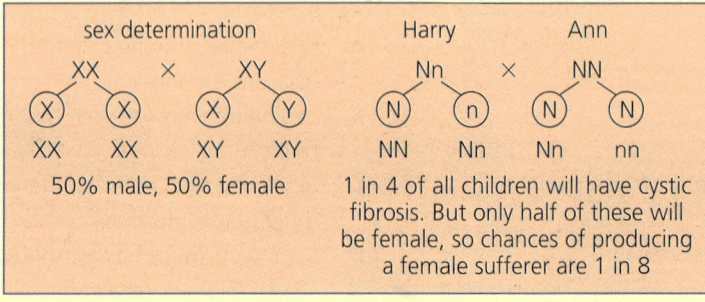

[4]

Chemical patterns

This module is about patterns that can be seen in

- the structure of atoms;
- the arrangement of elements in the periodic table;
- properties of the group of elements called halogens;
- the rates of chemical reactions.

Atoms and elements in the periodic table

You need to know •

✔ Atoms are made up from a nucleus containing protons and neutrons, with electrons moving around the nucleus in shells.

✔ approximately 100 elements are arranged in order of increasing atomic number in the periodic table;

✔ metals are on the left-hand side of the periodic pable and non-metals are on the right;

✔ elements in the same family have similar properties and are found in the same vertical column or group;

✔ examples of families are the alkali metals (Group 1), halogens (Group 7) and noble gases (Group 8);

✔ elements in the same group show a gradual change in properties from the top of the group to the bottom;

✔ there is a link between the arrangement of electrons around the nucleus of an atom of an element and its position in the periodic pable;

✔ the number of electrons in the outer shell of an atom is the same as the group number of the element in the periodic table.

Particles in atoms • • • • • • • • • • • • • •

Three different types of particle make up all atoms – protons, neutrons and electrons. The table below gives information about these three particles.

Atoms are the smallest part of an element that can exist alone.

Particle	Mass	Charge
proton, p	1 unit	+1
neutron, n	1 unit	0
electron, e	negligible	−1

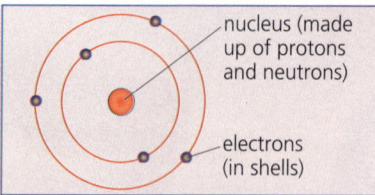

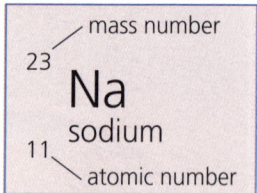

In atoms the protons and neutrons are in the **nucleus**. The protons and neutrons are tightly packed together. Electrons are arranged in shells around the nucleus. The nucleus of an atom is positively charged. Each atom has a **mass number** and an **atomic number**.

The atomic number is the number of protons in the nucleus of the atom.

Structure of the periodic table • • • • • • •

Elements are arranged in the **periodic table** in order of increasing **atomic number**. The horizontal rows of elements are called **periods**. Elements in the same family with similar properties are found in the same vertical column or **group**.

Examples of three chemical families are:

■ the alkali metals (in Group 1);

■ the halogens (in Group 7);

■ the noble gases (in Group 8).

Within each group there is a gradual change in physical and chemical properties.

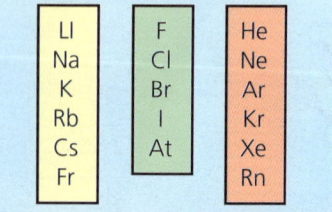

The bold line on the periodic table in the diagram below divides **metals** on the left-hand side of the line from **non-metals** on the right-hand side.

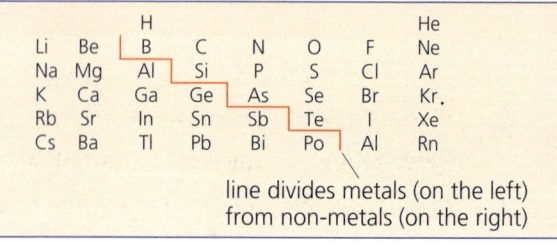

line divides metals (on the left) from non-metals (on the right)

There is a pattern in the arrangement of electrons of the elements in the periodic table. The diagram below shows the arrangement of electrons in atoms of the first 20 elements.

There is a pattern in the arrangement of electrons of the elements in the periodic table.

The number of electrons in the outer shell of an atom of an element is the same as the group number of the element in the periodic table. For example, carbon is in Group 4 and so has four electrons in the outer shell.

Elements in a group of the periodic table have similar properties. The physical and chemical properties of the elements change gradually down a group.

This pattern in properties can be seen with the alkali metals in Group 1, noble gases in Group 8 and halogens in Group 7.

A copy of the periodic table can be found on page 33.

There is one exception – helium is in Group 8 but has two electrons in the outer shell.

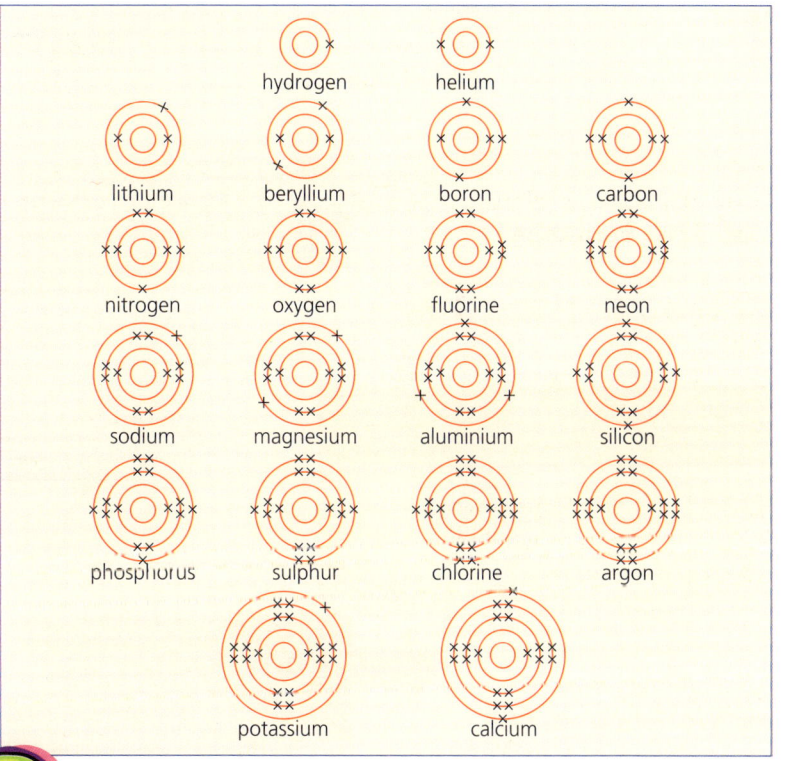

Questions

1 What are the two particles present in the nucleus of an atom?

2 Which particle has the smallest mass?

3 Which particle is not in the nucleus of an atom?

4 Look at the periodic table on page 33. Which element is in Period 3 and Group 5?

5 Is the element you found in Question 4 a metal or a non-metal?

6 Which of the elements below is in the same group as sodium?

   potassium   calcium   bromine   argon

7 How many electrons are there in the outer shell of a sulphur atom?

The properties of the halogens

You need to know •

✔ fluorine, chlorine, bromine and iodine belong to the halogen family;

✔ patterns in colour, melting points, boiling points and physical states of the halogens;

✔ the reactions of sodium and iron with chlorine;

✔ the differences in reactivity of the halogens;

✔ the uses of halogens.

The halogen family • • • • • • • • • • • •

The halogens are a family of similar elements in Group 7 of the periodic table. These elements are:

fluorine	F
chlorine	Cl
bromine	Br
iodine	I

Patterns in properties of the halogens • •

The table below shows the colour, melting point, boiling point and physical state at room temperature and atmospheric pressure of the three most common halogens.

Halogen	Colour	Melting point (°C)	Boiling point (°C)	State at room temperature
chlorine	green	−101	−35	gas
bromine	red-brown	−7	58	liquid
iodine	grey	114	183	solid

As you go down the group from chlorine the halogens become darker in colour. The melting points and boiling points increase down the group. The state changes from gas to solid.

The word halogen means salt producer. The salt sodium chloride is just called salt.

Reactions of sodium and iron with chlorine •

Chlorine is a very reactive halogen. It reacts with metals to form **salts**.

Sodium burns in **chlorine** to form solid **sodium chloride**:

$$2Na + Cl_2 \rightarrow 2NaCl$$

Iron reacts with **chlorine** gas to produce **iron(III) chloride**, a red-brown solid.

$$2Fe + 3Cl_2 \rightarrow 2FeCl_3$$

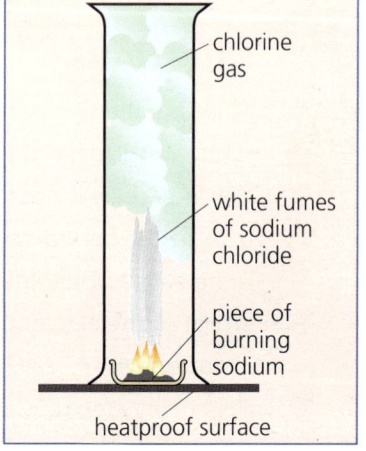

chlorine gas

white fumes of sodium chloride

piece of burning sodium

heatproof surface

Differences in reactivity of the halogens

The order of reactivity of the three halogens is chlorine (most reactive) > bromine > iodine (least reactive).

Displacement reactions show the differences in reactivity of the halogens.

Potassium and iodine combine in the compound potassium iodide. Chlorine gas added to potassium iodide solution causes a **displacement reaction**.

Chlorine, more reactive than iodine, replaces it in the salt **potassium iodide**:

$$Cl_2 + 2KI \rightarrow 2KCl + I_2$$

This reaction can be written as an **ionic equation**:

$$Cl_2 + 2I^- \rightarrow 2Cl^- + I_2$$

Bromine does not react with potassium chloride because bromine is less reactive than chlorine.

> The reactivity of the halogens decreases down Group 7.

> In an ionic equation, there must be the same number of atoms of each element on each side. Also, the sum of the charges on both sides must be the same.

> A reaction only takes place if the halogen added is more reactive than the halogen already present.

Uses of halogens

The table below summarises the common uses of two halogens – chlorine and iodine.

> Fluorine is a very reactive gas but fluoride added to water can reduce tooth decay.

Halogen	Use	Reason for use
chlorine	purifying water bleaching	kills bacteria that may be harmful household bleach contains chlorine as the active bleaching agent
iodine	as a solution in alcohol to kill bacteria	antiseptic for cuts on the skin

Questions

1. Which halogen is a liquid at room temperature and atmospheric pressure?
2. What is formed when chlorine is bubbled through potassium bromide solution?
3. What type of reaction is taking place in Question 2?
4. Which halogen is added to water to kill bacteria?
5. Write a symbol equation and an ionic equation for the reaction of chlorine with potassium bromide.

Rates of reaction

Reactions take place when particles of the reactants collide. The rate of reaction is increased when there are more frequent collisions or when there is greater energy on the collisions.

Rates of reactions with enzymes vary with changes in temperature and pH.

Chemical reactions take place at very different rates. For example, if a lighted splint is put into a test tube of hydrogen gas, the gas burns with a squeaky pop. This **explosion** is a very, very fast reaction and is finished in a tiny fraction of a second. This is a **very fast reaction**.

A limestone building will react with acidic gases in the atmosphere, but this reaction may take hundreds of years. This is a **very slow reaction**. The equation for the reaction is

$$Na_2S_2O_3 + 2HCl \rightarrow 2NaCl + SO_2 + H_2O + S$$

Most chemical reactions take place at rates between these two examples. If you want to study the rate of a chemical reaction it must not be too fast or too slow.

A very fast reaction

lighted splint

burns with squeaky pop

test tube filled with hydrogen

Remember that a fast reaction takes only a short time and a slow reaction takes a long time.

Effect of temperature on the rate of a reaction •

Increasing temperature increases the rate of a reaction. For example, we know that milk turns sour more quickly when left outside the refrigerator, especially in summer. The reaction souring the milk goes faster at warmer temperatures.

When solutions of sodium thiosulphate and dilute hydrochloric acid are mixed, a reaction takes place within a minute or so. One of the products is sulphur and this forms as a creamy precipitate.

The reaction can take place in a beaker placed on a piece of paper with a cross on it. The solutions are mixed and the time taken for the cross to disappear from view is noted.

If several reactions are carried out with equal concentrations of sodium thiosulphate and dilute hydrochloric acid but at different temperatures, it can be seen that as the temperature increases the time taken for the cross to disappear decreases.

sodium thiosulphate + hydrochloric acid

piece of paper with cross on it

Temperature (°C)	Time taken for cross to disappear (s)
20	200
30	120
45	50
55	18

Effect of concentration on the rate of a reaction

Magnesium and dilute hydrochloric acid react together to produce magnesium chloride and hydrogen gas. Magnesium dissolves in the solution.

> magnesium + hydrochloric acid → magnesium chloride + hydrogen

> $Mg(s) + 2HCl(aq) → MgCl_2(aq) + H_2(g)$

You can use state symbols in equations – (s) means solid, (aq) means dissolved in water, (l) means liquid, (g) means gas.

If the reaction is carried out using equal masses of magnesium but with different concentrations of dilute hydrochloric acid, a pattern can be seen in the results.

As the concentration of dilute hydrochloric acid increases, the time taken for the magnesium to disappear decreases – the reaction gets faster.

Concentration (g/dm^3)	Time taken for magnesium to disappear (s)
10	65
20	48
25	38
30	30
40	18

Effect of particle size on the rate of a reaction

In coal mines, coal dust mixed with air can explode. There is no risk of coal exploding when it is in lumps – in fact you might have difficulty getting lumps of coal to burn.

Calcium carbonate can exist in different forms. We can compare the reactions of a lump of calcium carbonate and an equal mass of calcium carbonate powder with equal volumes of dilute hydrochloric acid of the same concentration:

> calcium carbonate + hydrochloric acid → calcium chloride + water + carbon dioxide

> $CaCO_3(s) + 2HCl(aq) → CaCl_2(aq) + H_2O(l) + CO_2(g)$

We can measure the volume of gas produced every 30 seconds. The graph in the diagram shows the results of such an experiment.

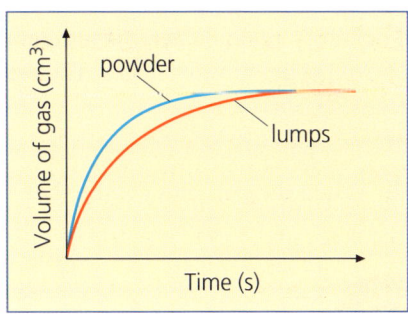

You can see that the same volume of gas is produced in each case because the same quantities of chemicals were used but the time taken is much less with powder – the reaction with the powder is much faster.

Using a catalyst

Some types of filler used to repair damage to car bodies comes in two tubes. When the contents of the two tubes are mixed the filler hardens. One tube contains a **catalyst**.

A catalyst is a substance that changes the rate of a reaction. Usually a catalyst speeds up a chemical reaction but there are some catalysts (called inhibitors) that slow reactions down. Only a small amount of catalyst is usually needed because it is not used up in the reaction.

continued ⟶

Hydrogen peroxide is a colourless liquid. It decomposes slowly without a catalyst to form oxygen gas and water. If manganese(IV) oxide is added to hydrogen peroxide the reaction is much faster.

Enzymes are biological *catalysts*. They work only under certain conditions. For example, most enzymes work best around 35–40°C. Above this temperature the structure of the proteins in the enzyme is said to be *denatured*.

Enzymes also work best over a specific range of pH.

More about enzymes will be found in Module 4.

Explaining the rate of a reaction using particles •

Substances are made up of **particles**. For a reaction to take place particles have to collide. A reaction gets faster when

- collisions take place more often;
- the colliding particles have more energy, e.g. when they are moving faster.

Increasing concentration
Look at A and B in the diagram below. Both boxes contain black and white particles.

Imagine that in both boxes the particles are moving in all directions at the same speed. A reaction takes place whenever the particles collide.

In B the particles are closer together, i.e. the reacting substances are more concentrated. There will be more collisions between black and white in B and therefore a faster reaction.

Increasing energy of collisions
Look at A again. When the temperature increases, the black and white particles move faster. The reaction then becomes faster, for two reasons:

- The particles collide more often.
- When the particles collide they have more energy.

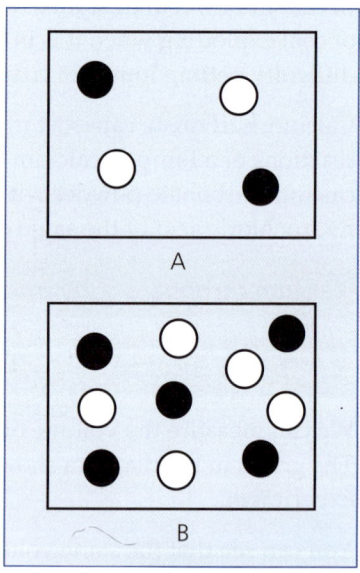

A

B

Questions

1   What name is given to a substance that speeds up a chemical reaction without being used up?

2   When a reaction is faster, does it take a shorter or longer time?

3   What name is given to a biological catalyst?

4   Why does powdered calcium carbonate react faster with hydrochloric acid than a lump of calcium carbonate?

group
numbers

0 (or 8)

1	2												3	4	5	6	7	He 2 helium

Li
3 lithium

Be
4 beryllium

Na
11 sodium

Mg
12 magnesium

K
19 potassium

Ca
20 calcium

Sc
21 scandium

Ti
22 titanium

V
23 vanadium

Cr
24 chromium

Mn
25 manganese

Fe
26 iron

Co
27 cobalt

Ni
28 nickel

Cu
29 copper

Zn
30 zinc

Rb
37 rubidium

Sr
38 strontium

Y
39 yttrium

Zr
40 zirconium

Nb
41 niobium

Mo
42 molybdenum

Tc
43 technetium

Ru
44 ruthenium

Rh
45 rhodium

Pd
46 palladium

Ag
47 silver

Cd
48 cadmium

Cs
55 caesium

Ba
56 barium

La
57 lanthanum

Hf
72 hafnium

Ta
73 tantalum

W
74 tungsten

Re
75 rhenium

Os
76 osmium

Ir
77 iridium

Pt
78 platinum

Au
79 gold

Hg
80 mercury

B
5 boron

C
6 carbon

N
7 nitrogen

O
8 oxygen

F
9 fluorine

Ne
10 neon

Al
13 aluminium

Si
14 silicon

P
15 phosphorous

S
16 sulphur

Cl
17 chlorine

Ar
18 argon

Ga
31 gallium

Ge
32 germanium

As
33 arsenic

Se
34 selenium

Br
35 bromine

Kr
36 krypton

In
49 indium

Sn
50 tin

Sb
51 antimony

Te
52 tellurium

I
53 iodine

Xe
54 xenon

Tl
81 thallium

Pb
82 lead

Bi
83 bismuth

Po
84 polonium

At
85 astatine

Rn
86 radon

H
1 hydrogen

⬛ = metals

⬛ = non-metals

⬛ = semi-metal

← metals

non-metals →

Practice module test

You will have 17 minutes to answer these questions

Use the graphs below to answer Questions 1 and 2.

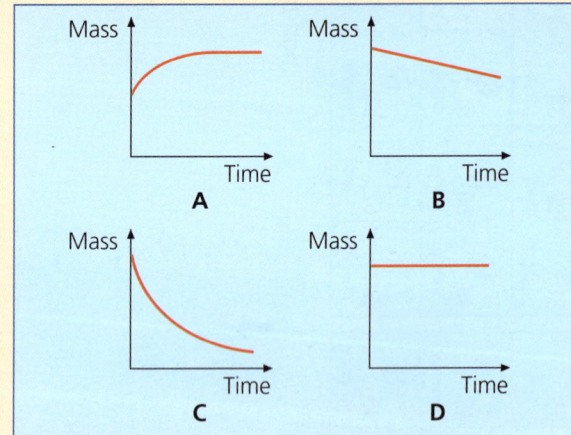

1 Which graph (**A**, **B**, **C** or **D**) shows the mass of calcium carbonate powder with excess dilute hydrochloric acid during the reaction?

2 Manganese(IV) oxide is a catalyst for the decomposition of hydrogen peroxide. Which graph (**A**, **B**, **C** or **D**) shows the mass of manganese(IV) oxide during the reaction?

3 Sodium reacts with chlorine to form sodium chloride. The correct equation for this reaction is:

   A $Na + Cl_2 \rightarrow NaCl_2$
   B $Na + Cl \rightarrow NaCl$
   C $2Na + Cl_2 \rightarrow 2NaCl$
   D $2Na + 2Cl \rightarrow (NaCl)_2$

4 Lithium, sodium and potassium are:

   A alkali metals
   B halogens
   C noble gases
   D transition metals

5 A reaction between lumps of calcium carbonate can be speeded up by:

   A adding water to the hydrochloric acid
   B lowering the temperature
   C breaking the lumps into smaller pieces
   D using half the volume of hydrochloric acid

6 Helium, neon and argon are:

   A alkali metals
   B halogens
   C noble gases
   D transition metals

7 Bromine, chlorine and iodine are halogen elements. Which of the following is true?

   A they are in Group 1
   B they are in Group 2
   C they are in Group 7
   D they are in Group 8

8 Iron reacts with chlorine to form iron(III) chloride. Which of the following correctly completes the equation: $2Fe + 3Cl_2 \rightarrow$

   A $FeCl_3$
   B $2FeCl_3$
   C Fe_3Cl
   D $2Fe_3Cl$

9 Lithium and sodium are elements in the same group of the periodic table. They must:

   A have the same number of electrons in the outer shell
   B be found in the same ore
   C have the same reaction with cold water
   D form coloured compounds

10 Which of the following pairs of substances will react?

   A potassium chloride and bromine
   B potassium chloride and iodine
   C potassium bromide and iodine
   D potassium bromide and chlorine

11 A magnesium atom contains 12 electrons. How many electrons are there in the outer shell of a magnesium atom?

   A 1
   B 2
   C 3
   D 4

12 Chlorine, bromine and iodine are in the same group. Which of the following statements is true?

    A The melting points and boiling points decrease down the group

    B The elements are in the same state at room temperature and pressure

    C The reactivity increases down the group

    D They all contain the same number of electrons in the outer shell

Questions 13 and 14 are about the reaction taking place when chlorine is bubbled through potassium iodide solution.

13 The colour change that takes place is from:

    A colourless to brown

    B brown to colourless

    C brown to green

    D colourless to green

14 The ionic equation for the reaction is:

    A $Cl + I^- \rightarrow Cl^- + I$

    B $Cl_2 + I^- \rightarrow 2Cl^- + I$

    C $Cl_2 + 2I^- \rightarrow 2Cl^- + I_2$

    D $Cl_2 + 2I^- \rightarrow 2Cl^- + 2I$

15 Chlorine is **not** used for:

    A making household bleaches

    B killing bacteria in water before using it as tap water

    C precipitating metals from water

    D sterilising water in swimming baths

Questions 16 and 17. Use this part of the periodic table.

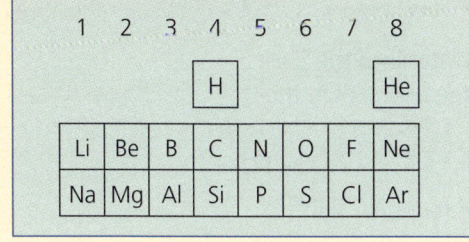

16 Which element has atoms with an electron arrangement of 2,8,4?

    A carbon

    B argon

    C silicon

    D magnesium

17 Which element has atoms with an electron arrangement of 2,8,8?

    A argon

    B helium

    C beryllium

    D magnesium

Questions 18 and 19. Look at the graph of volume of gas collected against time.

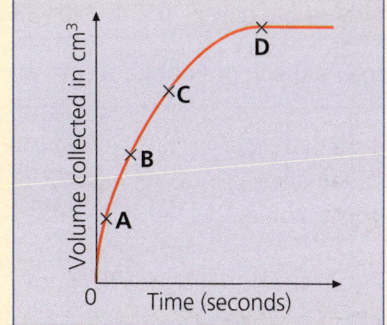

18 At which point (**A**, **B**, **C** or **D**) is the reaction fastest?

19 At which point (**A**, **B**, **C** or **D**) is half of the reaction completed?

20 For the three elements fluorine, lithium and magnesium, which of the following statements are true?

    A they are all in the same period

    B they are all in the same group

    C they are all metals

    D they all have two electrons in the innermost shell

Answers to these questions can be found on pages 143–147

Getting it right

1 Hydrogen peroxide decomposes into water and oxygen gas.

> **hydrogen peroxide → water + oxygen**

> **$2H_2O_2(aq) → 2H_2O(l) + O_2(g)$**

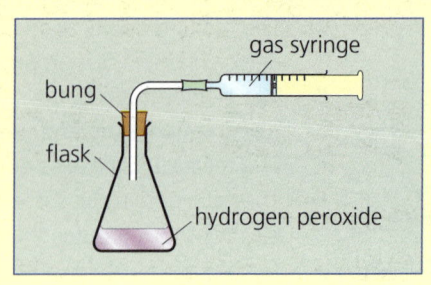

Bobby carries out an experiment using the apparatus in the diagram.

He puts 50 cm³ of hydrogen peroxide solution into the flask. He adds 0.5 g of manganese(IV) oxide and quickly pushes the bung into the flask. He measures the total volume of gas collected every minute. His results are shown in the table.

Time (minutes)	0	1	2	3	4	5	6	7	8	9	10
Volume of gas (cm³)	0	1	20	33	45	53	59	60	60	60	60

(a) Draw a graph of Bobby's results on the grid provided.

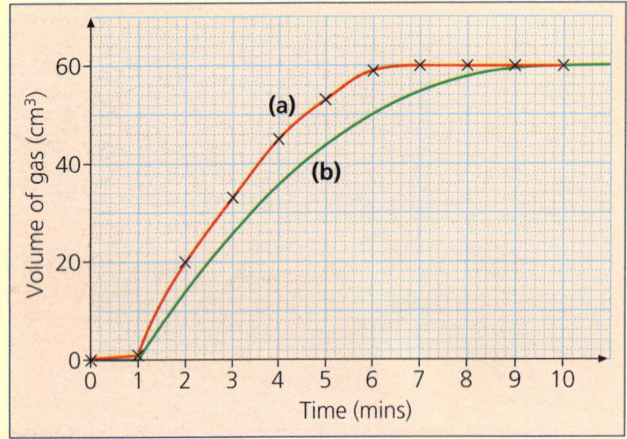

> *This graph is at standard demand. The axes and the scales are given. There are two marks for correct plotting and one for drawing the best line.*

[3]

(b) After how many minutes is the reaction finished?

7 minutes **[1]**

> *You can get this information from the table or the graph. The reaction is finished when no more gas is produced.*

(c) Bobby repeats the experiment but puts 50 cm³ of water into the flask before adding the manganese(IV) oxide. On the same grid sketch the graph he would get this time.

[2]

(d) Bobby wants to investigate whether the quantity of manganese(IV) oxide used affects the rate of reaction. Explain how he should do this.

Repeat using 5 different masses of manganese(IV) oxide. Same volume of hydrogen peroxide each time. Same concentration of hydrogen peroxide. Same temperature. Measure the volume of gas collected at intervals each time. Plot graphs – the steeper the graph the faster the reaction. **[5]**

Module 4

Chemistry in action

This module is about crude oil and the materials made from it.

Crude oil has formed in the Earth over millions of years.
After extraction, crude oil has to be refined by fractional distillation to make it into different fractions, with different boiling point ranges and different uses. Hydrocarbons produced from crude oil are commonly used as fuels.

There are different families of hydrocarbons. Alkanes are saturated hydrocarbons containing only single carbon–carbon bonds. Alkenes are unsaturated hydrocarbons containing at least one double carbon–carbon bond.

Plastics or polymers are made from crude oil and have a wide range of uses. They are made by the polymerisation of alkenes. Alkenes are produced by cracking long chain alkanes.

Crude oil

You need to know ••••••••••••••••••••••••••••••••••

✔ how crude oil was formed in the Earth;

✔ that crude oil is a mixture of hydrocarbons;

✔ that hydrocarbons are compounds of hydrogen and carbon only;

✔ that crude oil can be separated into fractions by fractional distillation and that each fraction has a different use;

✔ that the properties of a fraction are related to the size of its molecules.

How was crude oil formed? •••••••••••

Over millions of years, small animal organisms have died and decayed inside the Earth, forming a black sticky liquid that we call **crude oil**. This process required **high temperatures** (between 90 and 120ºC), **high pressure** and the **absence of oxygen**.
The crude oil is trapped in porous rocks between layers of impervious rock that prevent the oil from escaping.

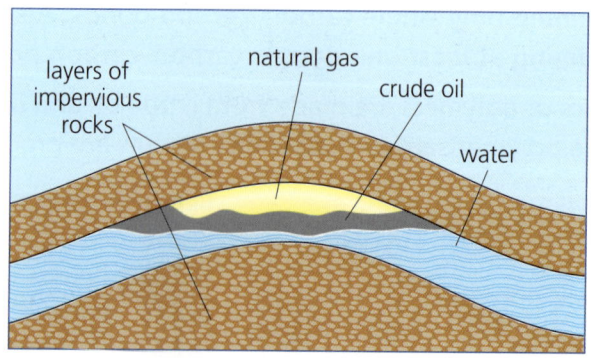

Crude oil trapped in rock layers in the Earth

Crude oil is a mixture of hydrocarbons

Crude oil is a complex mixture made up of different **compounds**. Most of these compounds are **hydrocarbons**. A hydrocarbon is a compound of carbon and hydrogen **only**. This mixture of hydrocarbons is of little use until the crude oil is refined.

Crude oil can be separated by fractional distillation •••••••••••••••••••••••••••••••

We use oil refineries to produce useful materials from crude oil by a process called **fractional distillation**.

Fractional distillation of crude oil involves separating the oil into different fractions, each having a different boiling point range.

The crude oil is boiled and the **vapour** formed enters a tall column. As the vapour passes up the tower, different compounds condense at different levels, depending upon their boiling point.

Compounds with high boiling points condense quickly, low down in the column. Compounds with low boiling points condense much further up the column where the temperature is lower.

> A compound is formed when atoms of different elements combine in fixed numbers, e.g. water (H_2O) is a compound of hydrogen and oxygen – two atoms of hydrogen and one of oxygen.

The diagram shows the fractional distillation of crude oil and some of the different fractions produced.

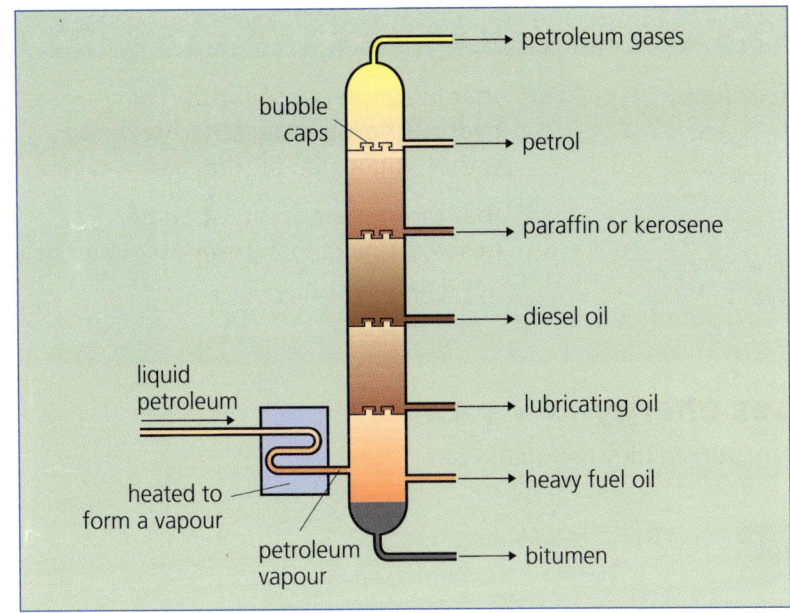

Fractional distillation column

Try to draw a fractionating column without looking at the book.

Fraction	Boiling point range (°C)	Number of carbon atoms	Uses
refinery gases	up to 40	1–4	fuel
petrol (gasoline)	40–140	5–10	fuel for cars
kerosene (paraffin)	140–180	8–12	fuel for aeroplanes
diesel oil	180–250	10–20	fuel for lorries and buses
fuel oil	250–340	20–40	fuel for central heating, ships
bitumen	over 340		tar for surfacing roads

The properties of each fraction depend upon the size of its molecules. The table shows that the boiling point increases as the molecules become larger (i.e. have more carbon atoms). Other properties of the various fractions also depend on the size of their molecules – for example, how easily they catch alight or how easily they pour.

Try to write down a list of five fractions obtained from crude oil, giving a use for each.

Questions

1   Which two elements are combined in a hydrocarbon?
2   What process is used in an oil refinery to make crude oil into useful products?
3   What three conditions were needed to turn small living organisms into crude oil?
4   What is the relationship between the size of a hydrocarbon molecule and where it condenses in a fractional distillation column?

Burning fuels

You need to know ••••••••••••••••••••••••••••••••••

✔ that burning (or combustion) of fuels releases energy;

✔ that the addition of oxygen makes this an **oxidation** reaction;

✔ that complete combustion of hydrocarbons produces carbon dioxide and water;

✔ that incomplete combustion of hydrocarbons produces carbon and carbon monoxide (a toxic gas);

✔ that incomplete combustion of hydrocarbons can happen in faulty gas appliances.

Burning of fuels releases energy ••••••••

Hydrocarbons are **fuels**. When they are burned they use up oxygen and release energy:

> **fuel + oxygen → products + energy**

A reaction where oxygen is added is an oxidation reaction, for example:

> **carbon + oxygen → carbon dioxide**
> $$C + O_2 \rightarrow CO_2$$

It is called oxidation because oxygen is added.

The products of combustion ••••••••••

The products of combustion (burning) depend upon the amount of air or oxygen present.

Complete combustion takes place when the fuel burns in a plentiful supply of air, i.e. there is an excess. It releases the maximum amount of energy. The products are **carbon dioxide** and **water**:

> **fuel + excess oxygen → carbon dioxide + water + energy**

When there is a limited amount of air or oxygen, we get **incomplete combustion**. Products include **carbon** (usually as soot) and **carbon monoxide**.

Carbon monoxide is a **toxic** gas. It is also colourless and odourless.

Incomplete combustion can be dangerous

Burning a hydrocarbon can produce poisonous carbon monoxide. This happens if the combustion is incomplete because not enough air reaches the burner. Every year about 50 people die of carbon monoxide poisoning in the UK.

Questions

1 What two substances are produced when natural gas burns in a plentiful supply of air?

2 What is the name of the poisonous gas produced when natural gas burns in a limited supply of air?

3 Write a symbol equation for the burning of methane where carbon (soot) is formed.

Plastics

You need to know •

✔ that methane, ethane, propane and butane are the first four members of a family of hydrocarbons;

✔ that cracking can break down long chain hydrocarbons, obtained by fractional distillation of crude oil, into more useful smaller molecules that often contain double bonds;

✔ that cracking takes place when hydrocarbon vapour is passed over a catalyst at high temperatures;

✔ that alkanes can be distinguished from alkenes by a test using bromine water – alkenes decolourise bromine water but alkanes do not;

✔ polymers are long chain molecules made up of large numbers of smaller molecules called monomers joined together – monomer molecules are unsaturated;

✔ polymers (the chemical name for plastics) have many uses;

✔ polymers do not easily rot away and so there are problems with their disposal.

Families of hydrocarbons • • • • • • • • • • •

Hydrocarbons exist in different families. One family is called **alkanes**. The simplest hydrocarbon in this family is **methane**. This has molecules with one carbon atom attached to four hydrogen atoms.

The next three members of the alkane family are:

C_2H_6	ethane
C_3H_8	propane
C_4H_{10}	butane

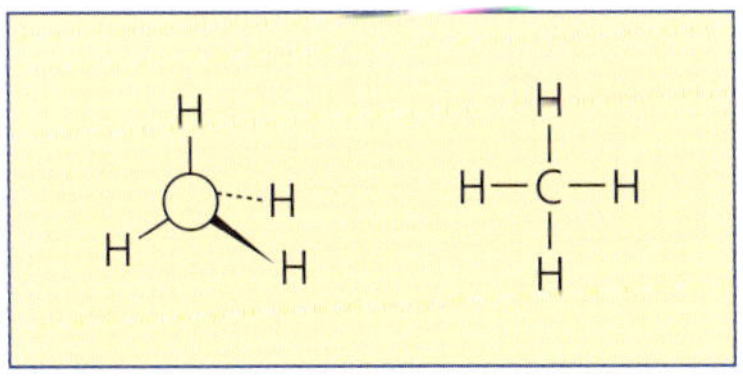

Structure of methane

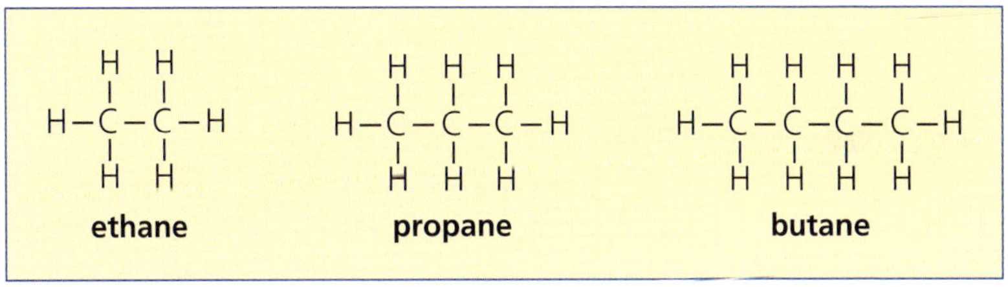

Structures of ethane, propane and butane

continued ⟶

Cracking can break down long chain hydrocarbons

There is not much demand for the long chain hydrocarbons formed from the fractional distillation of crude oil. It is profitable to break up these long chain hydrocarbons into smaller, more useful, molecules. This is done by a process called **cracking**. The vapour of the long chain molecules is passed over a catalyst at a high temperature, e.g.

> Cracking does not require high pressures and there must not be any oxygen present – otherwise the hydrocarbons would simply burn.

$$C_{10}H_{22} \rightarrow 4C_2H_4 + C_2H_6$$
decane   ethene   ethane

The small molecules produced by cracking include unsaturated molecules such as ethene and propene.

ethane   propene

> An unsaturated hydrocarbon molecule, e.g. ethene, contains one or more carbon–carbon double or triple bonds. A saturated hydrocarbon contains only single carbon–carbon bonds.

We can use a simple test to distinguish between alkanes and alkenes using bromine water. An alkene turns bromine water from red-brown to colourless – an alkane has no effect on bromine water.

ethane   bromine   dibromoethane

> Notice that when ethene reacts with bromine the carbon–carbon double bond is changed to a single bond. This is called an addition reaction.

Polymers and their applications

Fractions from crude oil are the starting materials for making polymers.

Polymers are long chain molecules, formed when thousands of small units called **monomers** join together.

A polymer chain

Monomer molecules are molecules containing carbon–carbon double bonds (unsaturated molecules). They can be joined together, usually using a catalyst, by a process of **addition polymerisation**.

Poly(ethene) is a polymer made by joining together many small ethene molecules, as shown in the diagram.

$$n \quad \underset{H}{\overset{H}{\diagdown}} C = C \underset{H}{\overset{H}{\diagup}} \quad \longrightarrow \quad \left[\begin{array}{c} H \;\; H \\ | \;\; | \\ C - C \\ | \;\; | \\ H \;\; H \end{array} \right]_n$$

This is an addition polymer because each carbon–carbon double bond becomes a single bond. A molecule cannot act as a monomer unless it contains a carbon–carbon bond.

Polymers have replaced materials such as metals and wood in many applications.

Polymer	Elements present	Uses
poly(ethene)	carbon and hydrogen	milk crates, cling film, dustbin bags, food storage boxes
poly(propene)	carbon and hydrogen	washing up bowls
poly(styrene)	carbon and hydrogen	flowerpots, ceiling tiles, plastic model kits
poly(chloroethene) or PVC	carbon, hydrogen and chlorine	insulation for electricity cables, guttering

Problems disposing of polymers • • • • • •

Polymers can cause litter problems because, unlike paper and cardboard, they do not rot away. They can be **recycled**, but this is often not economic.

When they are burned, polymers can form poisonous products such as carbon monoxide and hydrogen chloride (from PVC).

Questions

1   Which family of hydrocarbons includes methane, ethane and propane?

2   An alkane molecule contains n carbon atoms. How many hydrogen atoms does a molecule of this hydrocarbon contain?

3   Cyclohexane is a saturated hydrocarbon and cyclohexene is an unsaturated hydrocarbon. How could you distinguish between them?

4   Why is cracking an economic process?

Enzymes

Uses of enzymes at home and in industry

Enzymes in yeast ferment sugars to produce beer and wine. The enzymes act on sugar and turn it into ethanol and carbon dioxide:

A catalyst is a substance that speeds up the rate of a chemical reaction.

sugar + water → ethanol + carbon dioxide

$$C_{12}H_{22}O_{11} + H_2O \xrightarrow{\text{yeast}} 4C_2H_5OH + 4CO_2$$

A loaf of bread left to stand before baking will rise. The loaf rises as the above reaction produces bubbles of carbon dioxide, making the dough expand.

Biological washing powders are washing powders that contain enzymes. They are good at removing stains in cool water. The enzymes act on the stains and break them down.

Enzymes acting on milk produce cheese and yoghurt.

Injecting hard centred chocolates with the enzyme invertase makes soft centred chocolates.

Enzymes stop working at high temperatures ●

Enzyme reactions such as these work best at temperatures around 35°C. At higher temperatures the enzymes are broken down or **denatured**.

Students often write that enzymes are killed at high temperatures – this is incorrect.

Questions

1 Complete the equation for the fermentation of glucose by yeast:
$C_6H_{12}O_6 \rightarrow C_2H_5OH +$

2 Why do enzymes work well at about 35°C but do not work at 70°C

Types of chemical reactions

You need to know ●

✔ that chemical reactions can be classified according to their type;
✔ that reactions of acids with bases or alkalis to produce **salts** are called **neutralisation** reactions and can be used to produce salts;

✔ that a **thermal decomposition** reaction is where a substance is split up into simpler substances by heating;
✔ that limewater is produced by reacting calcium oxide with water;
✔ that calcium carbonate is used to produce glass, cement and iron.

There are many different types of chemical reaction. These include **combustion** (or burning), **oxidation** (where oxygen is added) and **reduction** (where oxidation is removed).

Neutralisation reactions ● ● ● ● ● ● ● ● ● ● ● ●

acid + base (or alkali) → salt + water

e.g.

$H_2SO_4 + 2NaOH \rightarrow Na_2SO_4 + 2H_2O$

Sometimes the salt produced can be added to soil as a **fertiliser**.

$2NH_4OH + H_2SO_4 \rightarrow (NH_4)_2SO_4 + 2H_2O$

Calcium oxide and calcium hydroxide are both alkaline and they can be added to soil to neutralise the excess soil acidity.

Plants need large quantities of the elements nitrogen, potassium and phosphorus to be healthy.

Thermal decomposition ● ● ● ● ● ● ● ●

The splitting up of a substance into simpler products is called **decomposition**. When this takes place by heating it is called **thermal decomposition**.

Two examples of thermal decomposition are the reactions of limestone (calcium carbonate) and copper carbonate:

$CaCO_3 \rightarrow CaO + CO_2$

$CuCO_3 \rightarrow CuO + CO_2$
green powder   black powder

Uses of calcium

Product	Other materials needed	Summary of process
glass	sodium carbonate and sand	melting the mixture together
cement	clay	heating limestone powder and clay and grinding the mixture to a powder
iron	iron ore and coke	heated in a furnace with blasts of hot air

Questions

1. Write down three materials made from calcium carbonate.
Ammonium sulphate is a salt produced from ammonia.
2. Write down the name of the acid used to make ammonium sulphate.
3. Suggest one use of ammonium sulphate.
4. Burning magnesium in oxygen produces magnesium oxide. Choose the best word in the list to describe this reaction.
   decomposition   oxidation   reduction

Practice module test

You will have 17 minutes to answer these questions

1 Crude oil was formed from small animals that lived in the sea. When they died their bodies were covered with layers of rock, which kept out air. What other conditions were needed to turn these into crude oil?

- **A** high pressures and low temperatures
- **B** low pressures and low temperatures
- **C** high pressures and high temperatures
- **D** low pressures and high temperatures

2 Which of the following word equations represents neutralisation?

- **A** lead carbonate → lead oxide + carbon dioxide
- **B** lead oxide + nitric acid → lead nitrate + water
- **C** lead nitrate → lead oxide + nitrogen dioxide + oxygen
- **D** lead oxide + carbon → lead + carbon monoxide

3 Which one of the following is a hydrocarbon?

- **A** C_6H_6
- **B** CO
- **C** C_2H_3Cl
- **D** $C_6H_{12}O_6$

4 Four fractions obtained from crude oil are bitumen, diesel, kerosene and petrol. In which list are the four fractions in order of increasing boiling point?

- **A** bitumen, diesel, petrol, kerosene
- **B** petrol, diesel, kerosene, bitumen
- **C** diesel, kerosene, petrol, bitumen
- **D** petrol, kerosene, diesel, bitumen

5 Which gas could be produced when poly(vinylchloride) or PVC is burned but not when poly(ethene) is burnt?

- **A** carbon dioxide
- **B** carbon monoxide
- **C** hydrogen chloride
- **D** water

6 An alkene contains

- **A** only single bonds
- **B** a double bond and single bonds
- **C** only double bonds
- **D** a single bond and double bonds

7 Which word equation represents fermentation?

- **A** sugar + oxygen → carbon dioxide + water
- **B** water + carbon dioxide → sugar + oxygen
- **C** sugar + water → ethanol + carbon dioxide
- **D** ethanol + oxygen → carbon dioxide + water

8 Crude oil was made from small marine organisms under the effect of:

- **A** high temperature, high pressure and lack of oxygen
- **B** low temperature, high pressure and lack of oxygen
- **C** high temperature, low pressure and presence of oxygen
- **D** low temperature, low pressure and presence of oxygen

9 Which gas is produced when a hydrocarbon burns in a limited supply of air but not in a plentiful supply of air?

- **A** carbon dioxide
- **B** carbon monoxide
- **C** oxygen
- **D** steam

10 Methane is a:

- **A** compound of carbon, hydrogen and oxygen
- **B** compound of carbon and hydrogen
- **C** mixture of carbon and hydrogen
- **D** mixture of carbon and water

11 The structure of a hydrocarbon is shown below. This hydrocarbon is:

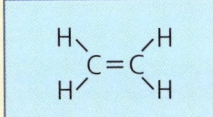

A an alkane
B a saturated hydrocarbon
C a solid at room temperature and atmospheric pressure
D called ethene

12 What is the molecular formula of an alkene containing three carbon atoms?

A C_3H_4
B C_3H_6
C C_3H_8
D C_3H_{10}

13 Poly(ethene) is produced from decane ($C_{10}H_{22}$) in a two-stage process:

stage 1   stage 2
decane → ethene → poly(ethene)

What occurs at stages 1 and 2?

	Stage 1	Stage 2
A	fractional distillation	polymerisation
B	oxidation	reduction
C	cracking	polymerisation
D	polymerisation	fractional distillation

14 Calcium hydroxide reacts with dilute hydrochloric acid according to the equation

$Ca(OH)_2 + xHCl → CaCl_2 + yH_2O$

What numbers should replace x and y if the equation is to be balanced?

A 2 and 2
B 2 and 1
C 1 and 2
D 4 and 2

15 Which of the following could be a monomer for an addition polymer?

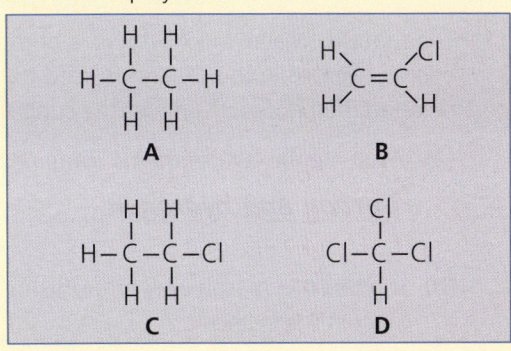

16 In the oil fractional distillation column, fractions collected near the top of the column:

A contain a mixture of small hydrocarbon molecules
B do not catch alight easily
C are difficult to pour
D contain compounds with a high boiling point

17 The equation for the burning of a hydrocarbon is excess air is

hydrocarbon $+ 3O_2 → 2CO_2 + 2H_2O$

What is the formula of the hydrocarbon?

A CH_2
B CH_4
C C_2H_4
D C_2H_6

18 The ionic equation for a neutralisation between hydrochloric acid and sodium hydroxide is

A $NaOH + HCl → NaCl + H_2O$
B $H^+ + OH^- → H_2O$
C $2H^+ + O_2 → H_2O$
D $NaOH + HCl → NaCl$

19 An alkene and an alkane can be distinguished using:

A Universal Indicator
B bromine water
C starch indicator
D Benedict's solution

20 In the test for an alkene, the solution turns from:

A brown to colourless
B colourless to brown
C red to blue
D green to orange

Answers to these questions can be found on pages 143–147

Getting it right

1   The rubbish collected in Anytown is burned in a large incinerator. The energy released in this process is used to provide hot water for neighbouring houses. Some of the rubbish consists of hydrocarbons.

(a) What are the two elements combined in all hydrocarbons.

Carbon and hydrogen. [1]

(b) Suggest one reason why rubbish should be burned in a plentiful supply of air in the incinerator.

To avoid carbon monoxide being formed, which is poisonous. [2]

In Brentown, the rubbish is buried underground in a landfill site. Over a number of years the rubbish breaks down and methane gas is produced. This gas is collected and burned on the site.

(c) Write a balanced equation for the burning of methane is excess air.

$CH_4 + 2O_2 \rightarrow CO_2 + 2H_2O$ [3]

> The three marks here are for the correct formulae on the left-hand side, correct formulae on the right-hand side and for balancing the equation

(d) Why is this gas collected and burned?

Mixtures of methane and air can explode. Collecting and burning it prevents this. [2]

(e) Suggest one reason why the disposal of rubbish in Anytown is better than in Brentown.

Energy released is used rather than wasted. [1]

> Apart from the risk of explosion, methane is a major greenhouse gas and it would further global warming.

2   Poly(ethene) is produced from high boiling point fractions from crude oil refining.

(a) Describe the stages in the production of poly(ethene).

The high boiling point fraction is broken down by the process of cracking; into smaller unsaturated molecules; using high temperature and a catalyst; ethene is the monomer; ethene is converted into poly(ethene) by addition polymerisation; using a catalyst. [6]

> There are seven points given in the answer but only five are needed to score 5 marks. The additional mark is for QWC. It is awarded if the candidate uses a suitable structure and style of writing..

(b) Complete the equation for the polymerisation of ethene.

$$n \begin{array}{c} H \\ \diagdown \\ H \diagup \end{array} C = C \begin{array}{c} H \\ \diagup \\ \diagdown H \end{array} \longrightarrow \left[\begin{array}{cc} H & H \\ | & | \\ C - C \\ | & | \\ H & H \end{array} \right]_n$$

[2]

Energy and electricity

5

This module is about electricity and energy resources.
The first topic in the module covers circuits and how
the current in a circuit is affected by the voltage and different circuit components.

Using electricity from the mains safely is the subject of the second topic in the module.
The final topic is concerned with generating electricity from different resources and how
to insulate buildings to conserve energy resources.

Circuits

You need to know •••••••••••••••••••••••••••••••

✔ about resistance and how it affects the current in a circuit;

✔ how the voltage across a component depends on its resistance and the current passing in it;

✔ how the current in a fixed resistor, a filament lamp and a diode depends on the voltage;

✔ how the resistance of a light-dependent resistor and a thermistor depend on the light intensity and temperature.

Direct and alternating currents •••••••

Electric **current** is a flow of charged particles. Cells and batteries cause a **direct current**, one that passes in the same direction all the time. The current from the mains supply is **alternating current** – it changes direction.

Any electric current that changes direction is an alternating current.

Units are important •••••••••••••••

The size of the current (measured in amps, A) in a circuit depends on the **resistance** (measured in ohms, Ω) and the **voltage** (measured in volts, V). Resistance is a measure of the opposition to current passing in a circuit. The greater the resistance, the smaller the current:

■ Increasing the resistance in a circuit decreases the current.

■ Increasing the voltage causes the current to increase.

The relationship between voltage, current and resistance is:

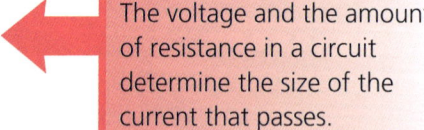

The voltage and the amount of resistance in a circuit determine the size of the current that passes.

voltage = current × resistance or $V = I \times R$

You need to know how the current in three common circuit components changes when the applied voltage is changed:

■ The current in a **fixed resistor** is proportional to the voltage. The resistance has a constant value.

■ The current in a **filament lamp** increases when the voltage is increased, but not in proportion – the resistance of the filament increases as it gets hotter.

■ The current in a **diode** can only pass in one direction (shown by the direction of the arrow on the circuit symbol); the diode only conducts when a minimum voltage (about 0.6 V) is reached – after this its resistance decreases as the voltage is increased.

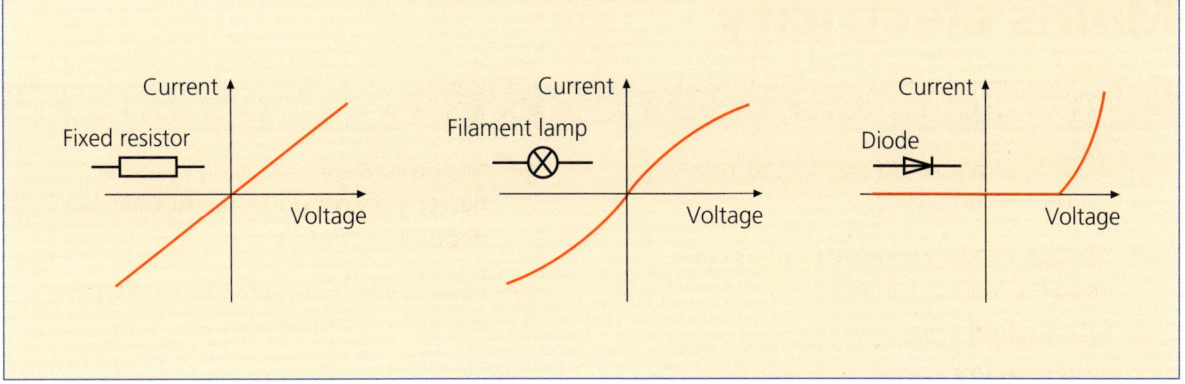

Light-dependent resistors (LDRs) are used in circuits that switch lights on automatically after dusk. **Thermistors** (temperature-dependent resistors) are used to switch the heaters in incubators on and off to maintain a constant temperature. The resistance of these components depends on the environmental conditions:

- the resistance of an LDR decreases when the light becomes brighter;
- the resistance of a thermistor decreases when its temperature increases.

The more energy being supplied from the environment to the component, the greater the current that passes.

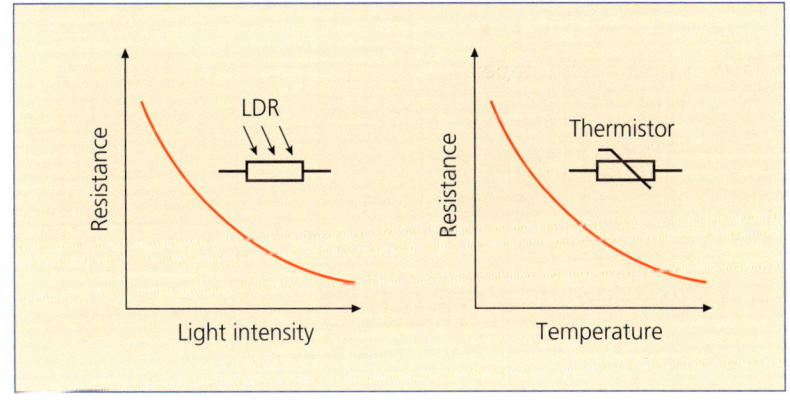

Questions

1 What is the difference between alternating and direct current?

2 Should a voltmeter be connected in series or in parallel with a component?

3 The current in a fixed resistance is 2.5 A when the voltage across it is 12.0 V. Calculate the resistance of the resistor.

4 Which device would you use to switch on a light automatically at night?

Mains electricity

You need to know •

✔ how appliances are connected safely to the mains supply;

✔ the job of each conductor in a mains lead;

✔ that electric current causes heating and how this is used in domestic appliances;

✔ the advantages of using residual current circuit breakers (RCCBs) instead of fuses;

✔ how to calculate the cost of energy from the mains supply.

The mains supply to a house and to a domestic appliance uses three conductors:

- The live wire is the high-voltage conductor. It carries the energy to the house or appliance.
- The neutral wire is there to complete the circuit.
- The earth wire is for safety. It prevents the casing of the appliance from becoming live if there is a break in the circuit.

The diagram shows the colours of the insulation used on the wires in a flexible cable connected to a plug.

Appliances that have a plastic casing and no metal exposed parts do not need an earth wire. These appliances are described as being **double insulated**, because there are two layers of insulation between the user and the live conductor. The plastic case cannot become live because it is not a conductor, so it provides an extra layer of insulation.

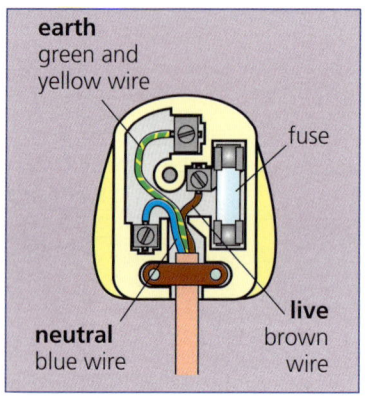

earth
green and yellow wire

fuse

live
brown wire

neutral
blue wire

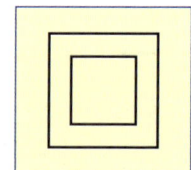

This symbol shows that an appliance is double insulated

The electricity supply entering a house passes through a consumer unit. In older houses each circuit is protected by a fuse, but in modern houses **residual current circuit breakers** (RCCBs) are used instead.

A fuse:

- consists of a thin wire that is heated by the current passing in it;
- melts and breaks the circuit when an excessive current passes;
- needs to be replaced by a new piece of wire of the correct rating when a fault has occurred and been repaired;
- is always fitted in the live conductor.

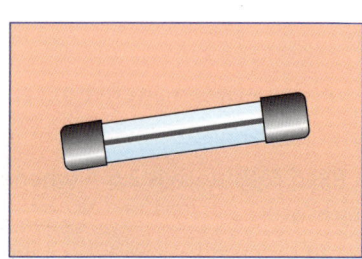

A cartridge fuse

An RCCB:

■ detects any difference between the currents in the live and neutral conductors, for example when a current passes to earth;

■ cuts off the live supply if this occurs;

■ can be reset easily by pressing a switch.

Domestic appliances transfer energy from electricity into **heat**, **light** and **movement** (including sound).

Whenever a current passes in a resistor, heat is produced. When an electric current passes in an appliance, there is always some heating because of the resistance of the wires.

Some appliances are designed to transfer energy from the electricity supply into heat. These include:

Electric bar heaters: the elements become hot and transfer energy by **electromagnetic radiation** to the surroundings.

Immersion heaters: these heat the water in the hot water tank. The hot water moves around the tank by **convection currents**.

Kettles, cookers and irons: these devices all use the heating effect of a current to raise the temperature of a heating element; in turn this heats the surroundings by transferring energy by convection currents and conduction. For example, an iron heats your clothes by transferring energy by conduction through the metal plate.

> Remember that sound is carried by the movement of particles.

this bar heater radiates energy

the water in this kettle is heated by convection currents

Electricity supply companies charge for the energy transferred from the mains supply to a home or workplace.

The energy is measured in **kilowatt-hours** (kW h), the amount of energy transferred by a 1 kW appliance in 1 hour.

The cost of this energy transfer can be calculated using the relationship:

> This relationship will always be given if required in module tests or exam.

cost = power in kW × time in h × cost of 1 kW h

Questions

1 In which conductor in a mains connecting lead is the fuse placed?

2 Which of the conductors in a mains connecting lead normally carries no current?

3 What causes an RCCB to cut off the electricity supply to a circuit?

Energy resources and transfer

You need to know •

✔ how electricity is generated;

✔ about the use of transformers in distributing electricity;

✔ the environmental implications of generating electricity using

renewable and non-renewable resources;

✔ how insulation uses trapped air to reduce the energy transfer between objects at different temperatures.

Electricity is generated by the rotation of a magnet within, or next to, a coil of wire. A small generator such as a bicycle dynamo uses a permanent magnet, but the generators in power stations use electromagnets that spin around inside thick copper conductors. The diagram on the right shows a bicycle dynamo.

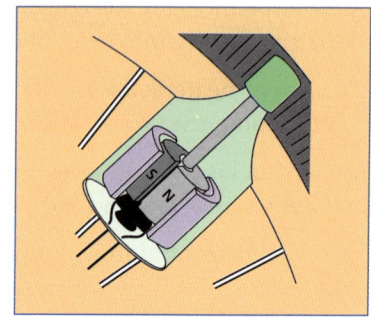

A bicycle dynamo

In a power station:

■ the electricity is generated at 25 000 V;

■ the voltage is increased (stepped up) using transformers to 400 000 V before being passed into the grid network;

■ electricity is transmitted using a combination of overhead and underground cables before the voltage is stepped down, by more transformers, for use by consumers.

Overhead cables:

■ are cheap to install and to operate because they are cooled by convection currents in the surrounding air;

■ are damaged easily in storms and windy weather;

■ can be repaired quickly because the damage can be seen easily;

■ are ugly and can spoil areas of natural beauty.

Underground cables:

■ are more expensive to install than overhead cables;

■ need a flow of coolant to remove excess heat;

■ are more costly to repair than overhead cables because damage cannot be detected easily.

Each wind turbine can generate 400 kw of electricity – enough for the needs of a small village of a few hundred people

Currently, the burning of fossil fuels such as coal and gas generates most of the UK's electricity supply. Fossil fuels are being used up rapidly and when they are burned they produce carbon dioxide (a greenhouse gas) and sulphur dioxide (which produces acid rain).

In the future we will need to use renewable sources of energy to generate electricity. These include:

■ Wind – large wind farms could be built offshore where the wind is reliable, there is plenty of room and the noise does not cause a nuisance.

There are plans to build large offshore wind farms off the east coast of England. These would be out of sight and hearing.

- Tides – a reliable, cost-effective generator of electricity from tidal flow has yet to be developed.
- Solar panels – these can be used to heat domestic water and to generate electricity directly.

The amount of polluting gases emitted from power stations can be reduced if there is less demand for electricity. This can be achieved by:

- using low-power appliances such as fluorescent lamps in place of filament lamps.
- improving the insulation of homes and workplaces.

Keeping the heat in

For most of the year, the temperature inside your house is kept at a higher level than the temperature outside, possibly by central heating. This results in a flow of energy from the inside of your house to the outside because of the difference in temperature (as shown in the diagram). The amount of energy flow can be reduced in several ways:

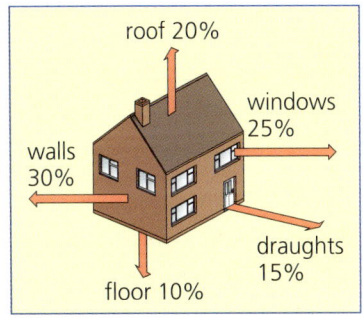

- **loft insulation** reduces the energy conducted through bedroom ceilings and convected through the air space in a loft.
- **double glazing** reduces the energy conducted through windows.
- **cavity wall insulation** stops energy transfer by convection currents between the inner and outer walls of a house.

Energy can still travel through an insulated cavity wall by conduction.

Each of these methods of insulation uses air as the insulator. Air is a very poor conductor of heat but it is very good at transferring energy through convection currents. If air is trapped so that it cannot form convection currents, it becomes a very good insulator. Air can be trapped in the following ways:

- Loft insulation usually consists of fibre glass, which traps air between the fibres.
- Double glazing is most effective when a thin layer of air is trapped between two sheets of glass; a thin layer of air cannot form convection currents because there is too much resistance to movement over the glass surfaces.
- Cavity wall insulation traps air in pockets of foam or mineral wool; this prevents convection currents in the cavity.

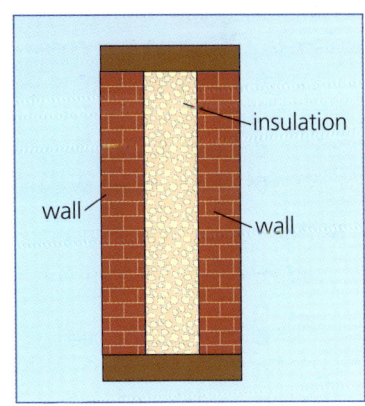

Insulating the cavity between an inner and an outer wall

Questions

1 How are electromagnets used to generate electricity in a power station?
2 What devices are used to change the voltage of mains electricity?
3 What type of fuel are coal, oil and gas?
4 Why is trapped air a good insulator?

Practice module test

You will have 20 minutes to answer these questions

1 Which part of a cycle dynamo rotates when it generates electricity?

    A the electromagnet

    B the permanent magnet

    C the temporary magnet

    D the coil of wire

2 Which line on the graph (**A**, **B**, **C** or **D**) shows a direct current?

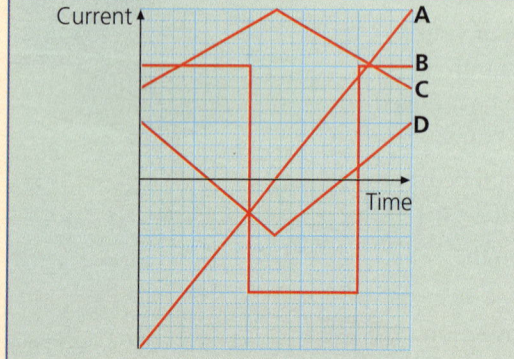

3 A diode:

    A allows current to pass in both directions at once

    B allows current to pass in one direction only

    C only allows alternating current to pass

    D does not allow direct current to pass

4 The colour of the insulation on the earth wire in a correctly wired plug is:

    A brown

    B brown and yellow

    C green

    D green and yellow

5 An appliance that is double insulated:

    A does not need a live wire

    B does not need a neutral wire

    C does not need an earth wire

    D does not need a fuse

6 A transformer can be used to:

    A generate a direct voltage

    B generate an alternating voltage

    C change the size of a direct voltage

    D change the size of an alternating voltage

Questions 7 to 9

Here are some methods used to insulate a home:

    A cavity wall insulation

    B double glazing

    C draught-proofing

    D loft insulation

7 Which of these methods reduces the energy lost through a bedroom ceiling?

8 Which of these reduces the energy lost through the walls?

9 Which of these reduces the energy lost through the windows?

10 The energy input to a wind generator is:

    A gravitational energy

    B kinetic energy

    C potential energy

    D pressure energy

Questions 11 and 12

Electricity is distributed at high voltage using both overhead and underground cables.

11 A disadvantage of using overhead cables is:

    A they need to have a coolant pumped through them

    B they are more expensive to install than underground cables

    C faults cannot be spotted easily

    D they are more prone to damage in adverse weather conditions

12 A disadvantage of using underground cables is:

    A they are more expensive to manufacture

    B they can be damaged by burrowing animals

    C they do not conduct electricity when the ground is cold

    D they crack up in freezing weather

13 The current in the circuit at the top of the next page can be increased by:

    A increasing the light intensity

    B decreasing the light intensity

    C increasing the temperature

    D decreasing the temperature

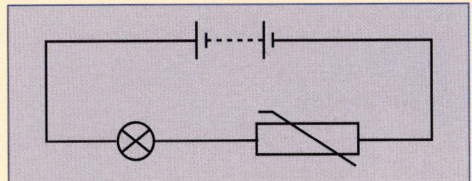

14 When a fuse "blows", it:

    A breaks the circuit in the live conductor

    B breaks the circuit in the neutral conductor

    C breaks the circuit in the earth conductor

    D breaks the circuit in all three conductors

15 A residual current circuit breaker is preferred to a fuse for some applications because it:

    A works at a higher voltage than a fuse

    B works at a higher current than a fuse

    C acts faster than a fuse

    D has less resistance than a fuse

16 Electricity is generated in a cycle dynamo by:

    A a permanent magnet oscillating inside a coil of wire

    B a permanent magnet rotating inside a coil of wire

    C an electromagnet oscillating inside a coil of wire

    D an electromagnet rotating inside a coil of wire

17 What is the main insulator in a double glazed window?

    A air

    B glass

    C plastic

    D a vacuum

18 In which circuits does the lamp light?

    A A and B

    B C and D

    C A and C

    D B and D

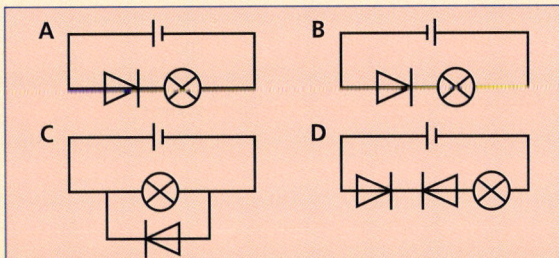

19 The resistance of a lamp filament is 6.0 Ω. When the voltage across the lamp is 12 V the current in the filament is:

    A 0.5 A

    B 2.0 A

    C 18 A

    D 72 A

20 A residual current circuit breaker breaks the circuit if:

    A there is more current in the live wire than in the earth wire

    B there is less current in the live wire than in the earth wire

    C there is more current in the neutral wire than in the live wire

    D there is less current in the neutral wire than in the live wire

21 An appliance that is double insulated:

    A does not need a live wire

    B does not need an earth wire

    C does not need a neutral wire

    D does not need a fuse

22 Air that is trapped cannot transfer energy by:

    A conduction

    B convection

    C radiation

    D any method

23 Transformers are used to increase the voltage of the electricity generated at a power station before it passes into the National Grid. The voltage is increased:

    A so that more power can be transmitted

    B so that less power is transmitted

    C so that power is transmitted at a low current

    D so that power is transmitted at a high current

24 Wood is a renewable source of energy. This means that:

    A wood can be burned over and over again

    B no more wood can be grown

    C wood takes too long to grow to provide new supplies

    D more wood can be grown within the lifetime of the Earth

Answers to these questions can be found on pages 143–147

Getting it right

(a) (i) In normal use, the current in a kettle element is 10 A. Which of the following fuses should be fitted to the plug?

    1 A   3 A   5 A   13 A

Explain why the others are not suitable.

> *13A, as the others would "blow" with a current*
> *of 10A.* **[2]**

(ii) What happens to a fuse when too great a current passes in it?

> *The wire melts and breaks the circuit.* **[2]**

(b) A kettle has a power of 2.4 kW. The cost of energy from electricity is 7p/kWh. Use the relationship:

    cost = power × time × cost in kW h

to calculate the cost of using the kettle for a total of 2 hours each day for a week.

> *cost = 2.4 x 14 x 7p = 235.2p* **[2]**

(c) The diagram shows a kettle that has a double casing, with a layer of air trapped between the inner and outer walls

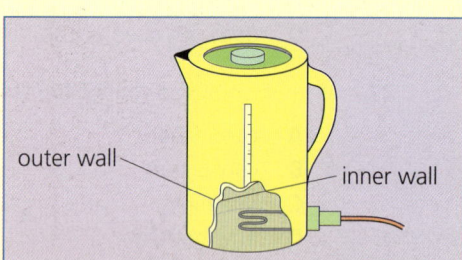

outer wall — inner wall

(i) Explain how this reduces the energy loss from the heated water.

> *The trapped air cannot form convection*
> *currents to transfer energy from the inner*
> *wall to the outer wall.*
> *Air is a poor conductor, so little energy is*
> *lost through conduction.* **[3]**

(ii) Suggest one other advantage of the kettle having a double outer casing.

> *The outer wall does not get hot, so the*
> *person using the kettle cannot get burned.* **[2]**

Usually an appliance should be fitted with a fuse that has a slightly higher rating than the current that passes when it operates normally.

When answering questions of this type, you do not have to change the final answer into pounds, but if you do, then make sure you do it correctly.

Where there are two marks available for the answer to a question, always make two separate points. In this case the two points are the increased safety and the reason for this.

Waves, space and atoms

6

The title of this module describes the three topics in it. The waves section concentrates on the use of the waves making up the electromagnetic spectrum. Electromagnetic waves are used to gather evidence and information about space, the subject of the second topic. The final topic is about radioactivity – its existence, benefits and dangers.

Waves

You need to know ● ● ● ● ● ● ● ● ● ● ● ● ● ● ● ● ● ●

✔ what the difference is between a longitudinal and a transverse wave, and how to describe both of them;

✔ the meaning of the terms **frequency**, **wavelength** and **amplitude**;

✔ the order of the electromagnetic

spectrum and how to describe the uses and hazards of each group of waves;

✔ about the causes and effects of refraction;

✔ about ultrasound and its uses.

There are two types of waves:

■ In a **longitudinal wave** the vibrations are parallel to the direction of wave travel.

■ In a **transverse wave** the vibrations are at right angles to the direction of wave travel.

Wave measurements that are common to both types of wave are:

■ Frequency (symbol f) – the number of waves occurring each second. Frequency is measured in Hertz (Hz).

■ Wavelength (symbol λ) – the length of one cycle of a wave. Wavelength is measured in metres (m).

■ Amplitude (symbol a) – the greatest displacement from the normal position. Amplitude is also measured in metres (m).

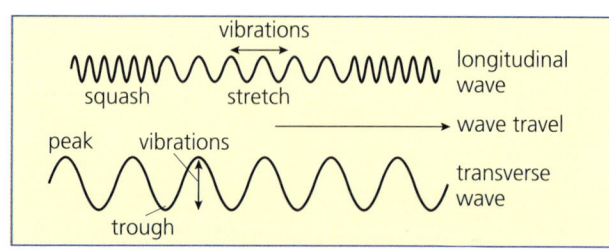

Vibrations in longitudinal and transverse waves

One cycle of a longitudinal wave is made up of a squash and a stretch. For a transverse wave the cycle is made up of a trough and a crest.

Electromagnetic waves such as light and radio waves are transverse waves – they all travel at the same speed in a vacuum. They cover a whole range of wavelengths and frequencies. The shorter the wavelength and higher the frequency, the more energy the wave carries.

High-energy waves can be very damaging to people and other animals if they are absorbed by cells.

The different groups of waves in the electromagnetic spectrum are shown in the table.

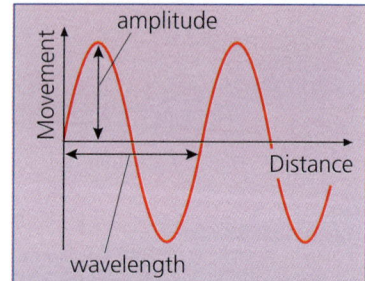

Measuring amplitude and wavelength in transverse waves

Wave group	Gamma rays	X-rays	Ultraviolet	Light	Infrared	Microwaves	Radio waves
Main use	Sterilising medical instruments and treating cancer	Looking inside the body and machines such as jet engines	Sun beds, fluorescent lamps and security marking	Seeing and photography	Remote controls and cooking such as grilling	Cooking and telephone and TV signals	Radio and TV broadcasting
Danger	Can cause cancer and damage to body cells and tissue	Can cause cancer and damage to body cells and tissue	Can cause skin cancer and damage to the eyes	Bright lights can damage the eyes	Causes skin burns	Causes heating of cells and tissue	None known

Electromagnetic waves are used to transmit information such as telephone calls, television and radio. These may be sent as either **analogue** or **digital** signals.

An analogue signal:

■ is continually changing and can have any value up to the maximum value.

A digital signal:

■ can only have certain values, usually 0 or 1;

■ gives clearer reception as any unwanted noise is easily removed.

Both light and infrared waves are used to send information in digital form through an **optical fibre**. Because there is little energy loss in the optical fibre, these signals can travel long distances in optical fibres before they need to be amplified.

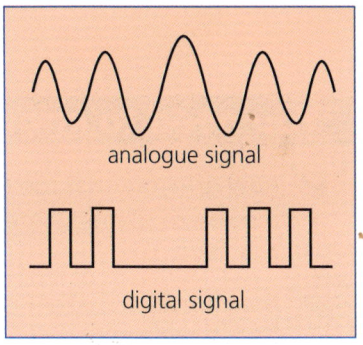

Analogue and digital signals

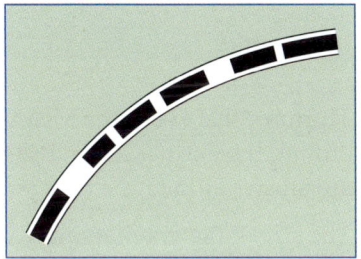

A digital signal travelling along an optical fibre

Light changing direction ● ● ● ● ● ● ● ● ● ● ●

Light and other electromagnetic waves normally travel in straight lines, but when they travel from one material into another their speed changes. This is known as **refraction** and it can cause a change in direction.

■ When light travels from glass into air it slows down. It speeds up again as it leaves.

■ Light travelling along a "normal" (a line drawn at right angles to the surface) does not change direction.

Sound is a longitudinal wave. It travels as a series of squashes and stretches. Human ears can detect sounds with frequencies in the range of 20 Hz to 20 000 Hz.

Sound waves with a frequency above 20 000 Hz are called **ultrasound.** Humans cannot hear ultrasound, although animals such as dogs and bats can detect some.
Ultrasound is used to:

■ produce computer images of the inside of the body – unlike X-rays, ultrasound does not damage cells and tissues so it is safe to use to examine a fetus, for example;

■ explore the seabed and locate shipwrecks and shoals of fish.

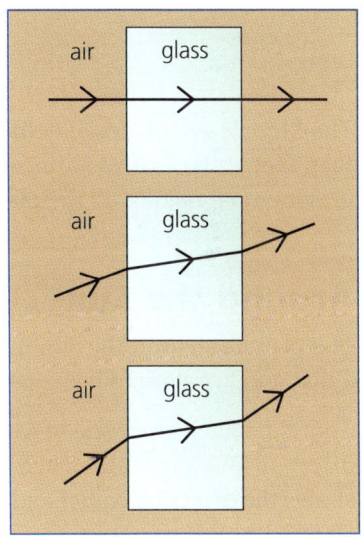

The effects of refraction

Questions

1 A ship sends a pulse of ultrasound to the seabed and detects the echo 0.84 s later. The speed of sound in water is 1500 m/s. How deep is the sea?

2 State one difference between sound and ultrasound.

3 Suggest why gamma rays can sterilise medical instruments but radio waves cannot.

4 Why is ultrasound preferred to X-rays for 'seeing' inside the body?

Space

The force that keeps the Moon and artificial satellites in orbit around the Earth is a **gravitational force**. Gravitational forces are always attractive – they pull objects towards each other.

> An artificial satellite is one that has been launched from Earth, for example a weather or communications satellite.

The size of the gravitational force acting on an object depends on its mass and the **gravitational field strength**, g.

The relationship between the force on an object, F, its mass, m, and the gravitational field strength is:

> **Force = mass × gravitational field strength or $F = m \times g$**

> *You need to be able to recall this relationship. It will not be given on exam or test papers.*

The Moon's gravitational field strength at its surface is much less than that at the Earth's surface. Jupiter has the strongest gravitational field of all the planets because it is very massive.

Around the Sun ● ● ● ● ● ● ● ● ● ● ● ● ● ● ● ● ● ●

The planets all orbit the Sun in the same direction. Their orbits form a disc and are close to being circular.

Comets, which are made of ice and rock, can orbit the Sun in any direction and any plane. Their orbits are elliptical. The diagram shows the orbit of a comet and how the force on the comet changes during its orbit.

Groups of millions of stars and their satellites that are held together by gravitational forces are called **galaxies**. The diagram shows the Milky Way galaxy and the position of our Solar System.

The number of galaxies that make up the whole **Universe** is so vast that they cannot be counted.

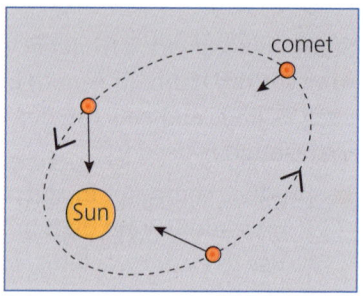

A comet's orbit around the sun

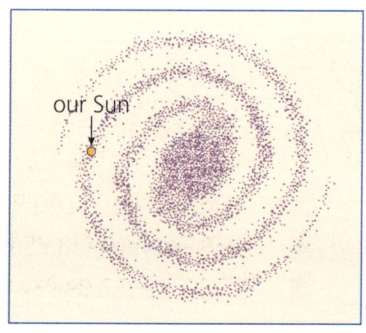

Extraterrestrial ● ● ● ● ● ● ● ● ● ● ● ● ●

Many scientists believe that somewhere there must be a planet with conditions similar to those on Earth, where some form of life exists.

Ways of looking for life on other planets include:

■ For near planets in our Solar System, such as Mars, we can send space probes to land on the planet. These probes carry out experiments on soil to search for evidence of primitive forms of life such as microbes.

■ Scanning the radio wave spectrum to try to detect transmissions from more distant planets.

Stars give out light as a result of energy released in nuclear reactions. Stars are not only being formed continually, but are also coming to the end of their existence as stars.

The stages in the life cycle of a small star like our Sun are:

■ gravitational forces cause a cloud of dust and gas to collapse;

■ as the cloud collapses, it becomes hot;

■ when the temperature is high enough for nuclear fusion to start, a Sun is born;

■ our Sun is in its **main sequence**, where the reaction that generates heat and light is the **fusion**, or joining together, of hydrogen nuclei to form helium nuclei;

■ after its main sequence, our Sun will expand and cool to form a **red giant**;

■ it will then contract and heat up again until it is a **white dwarf** and then cool to an invisible **black dwarf**.

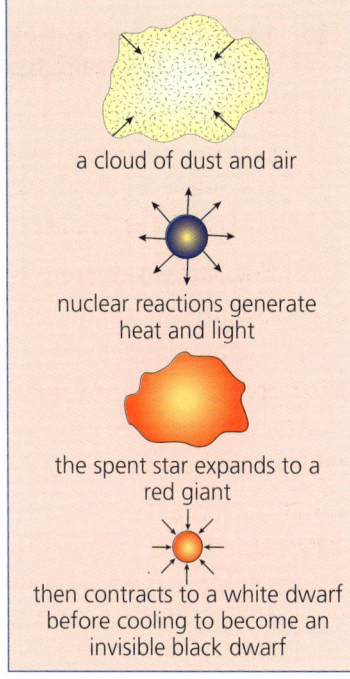

The stages in the life of a star

The Universe is thought to have begun with an enormous explosion, called the "**Big Bang**". Matter was created in this explosion, which has since formed into planets and stars.

Evidence for the "Big Bang" theory includes:

■ Red shift – the light received from most galaxies has its wavelength shifted towards the red end of the spectrum, showing that the galaxies are moving away from us.

■ Microwave radiation – the whole of space is filled with microwave radiation left over from the original explosion.

You have probably observed that the noise from an ambulance or aircraft seems to become lower in pitch as it travels away from you – this is an example of red shift in sound waves.

The Universe is known to be expanding. In the future it could:

■ continue to expand if there is not enough mass for gravitational forces to halt this;

■ reach and maintain a steady size;

■ stop expanding and contract if there is enough mass for the gravitational forces to be strong enough.

Questions

1 The Sun's pull on Jupiter is greater than the Sun's pull on the Earth, even though Jupiter is further away from the Sun. Suggest the reason for this.

2 State two ways in which the force acting on a comet changes during its orbit.

3 If light from a galaxy shows "blue shift", what does it show about the galaxy?

Atoms

You need to know ●

✔ how to describe the atomic structure of an isotope from its symbol;

✔ how to describe alpha, beta and gamma radiation, their properties and some common uses;

✔ the meaning of background radiation and its sources;

✔ about the dangers of ionising radiation and the difficulties in disposing of radioactive waste.

The atomic nucleus contains two types of particle:

■ Protons have a positive charge – the number of protons determines the element.

■ Neutrons have the same mass as protons but are not charged. Atoms of the same element can have different numbers of neutrons.

Atoms are represented by a symbol that tells you about the tructure of the nucleus. For example, the symbol for the most common form of carbon is $^{12}_{6}C$. The lower number, 6, is the atomic number and represents the number of protons in the nucleus. This identifies the element as carbon. The upper number is the total number of protons and neutrons (together called nucleons).

> The number of protons is the same as the position of the element in the periodic table.

There are other forms of carbon, called isotopes. These include $^{14}_{6}C$ and $^{11}_{6}C$.

The nuclei of carbon-14, , $^{14}_{6}C$ and carbon-11, $^{11}_{6}C$, are unstable. This means that:

■ The nucleus can break down, giving out one or more radioactive emissions.

■ This process is unpredictable and random.

■ When the nucleus breaks down, a more stable nucleus of a different element is formed.

> Different isotopes of an element have the same number of protons but different numbers of neutrons. Work out the numbers of protons and neutrons in $^{14}_{6}C$ and $^{11}_{6}C$.

Properties of the three main types of radioactive emission are shown in the table.

Radioactive	Nature	Charge	Mass	Penetration
Alpha particle	Two neutrons and two protons	+2	4 × the mass of a proton	Absorbed by a sheet of paper or a few cm of air
Beta particle	Fast-moving electron	−1	1/2000 the mass of a proton	Absorbed by 3 mm of aluminium
Gamma ray	Short-wavelength electromagnetic radiation	0	0	Intensity is reduced by several cm of lead

> It is not possible to completely absorb gamma radiation – its intensity can only be reduced.

Some common uses of radioactivity include:

■ Alpha particles are used in smoke alarms – the smoke absorbs the alpha particles so they cannot ionise the air.

■ Beta particles are used to control the thickness of sheet materials such as aluminium foil.

■ Gamma rays are used to sterilise food and medical instruments.

Gamma rays are used for sterilising because they kill microbes.

We live in a radioactive world – we are constantly being bombarded with radiation that is called **background radiation**. The chart shows the sources of background radiation in the UK.

Radiation from the ground varies according to where you live. Radiation is much higher in places where there is granite because this emits radioactive radon gas.

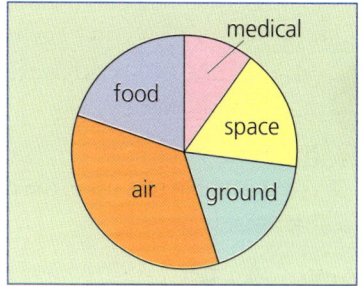

Exposure to radiation can be damaging to humans. It can:

■ cause mutations – changes to the DNA of sex cells can result in the birth of deformed offspring;

■ damage cells and tissue, which results in cancer.

Because of this, we need to take care when disposing of radioactive waste from hospitals and nuclear power stations.

Radioactive waste can emit harmful radiation for thousands of years. It needs to be placed in sites where the radiation cannot reach and damage people. Burying it underground is one possibility but there are problems with doing this:

■ There could be changes in the geology that may expose the material or cause the containers to crack.

■ Radioactive material could enter the water supply.

■ The waste needs to be buried in very deep chambers so that the emissions are absorbed by the ground.

■ There is a danger of making the soil radioactive.

The problems of disposing of radioactive waste have resulted in a decline in the number of nuclear power stations in the UK over recent years.

Questions

1 Oxygen-15 and oxygen-16 are isotopes of oxygen. In what way are they the same and in what way are they different?

2 Smoke alarms use sources of alpha radiation. Explain why these sources are not harmful to people in the same room.

3 Radioactive decay is random. What does this mean?

Practice module test

You will have 20 minutes to answer these questions

1 The frequency of a wave is:

    A the time taken to complete one cycle
    B the length of one cycle
    C the number of waves that occur each day
    D the number of waves that occur each second

2 In the visible spectrum, the colour that has the longest wavelength is:

    A red
    B orange
    C green
    D blue

3 When light is refracted, changes occur to its:

    A speed and wavelength
    B speed and frequency
    C frequency and direction
    D frequency and wavelength

4 Fluorescent lamps emit:

    A X-rays
    B gamma rays
    C ultraviolet radiation
    D radio waves

5 Light slows down as it travels from air into glass. This is known as:

    A deflection
    B diffraction
    C reflection
    D refraction

6 A moon is:

    A a planet
    B a satellite of a planet
    C a satellite of a star
    D the centre of a solar system

7 $^{14}_{6}C$ has:

    A 6 neutrons in its nucleus
    B 14 neutrons in its nucleus
    C 6 protons in its nucleus
    D 14 protons in its nucleus

8 The gravitational field strength at the surface of the Earth is 10 N/kg.

That at the surface of Jupiter is 24 N/kg.

A person weighs 600 N on Earth. The person's weight on Jupiter would be:

    A 60 kg
    B 1440 kg
    C 60 N
    D 1440 N

9 Gravitational forces:

    A only act between objects that are charged
    B only act between objects that are magnetic
    C are always attractive
    D are always repulsive

10 The orbits of comets:

    A are circular
    B are elliptical
    C are always in the same plane
    D are always in the same direction

11 The nucleus of an atom represented by the symbol $^{23}_{11}Na$ has:

    A 11 protons
    B 12 protons
    C 11 nucleons
    D 12 nucleons

12 Of the three main types of radioactive emission, alpha radiation:

    A has the smallest charge
    B has the smallest mass
    C has the lowest penetration
    D causes the least ionisation

13 Light travels at a speed of 3.0×10^8 m/s in a vacuum. The speed of light in glass could be:

A 2.2×10^8 m/s
B 3.0×10^8 m/s
C 3.2×10^8 m/s
D 3.5×10^8 m/s

14 The energy carried by an electromagnetic wave depends on its:

A amplitude
B frequency
C speed
D wavelength

15 X-rays are used to examine a bone fracture because they:

A are absorbed by flesh but not by bone
B are absorbed by bone but not by flesh
C are absorbed by both flesh and bone
D are absorbed by neither flesh nor bone

16 A compression wave with a frequency of 25 000 Hz is known as:

A infrared
B sound
C ultrasound
D ultraviolet

17 Which of these is an example of "red shift"?

A the sound from an aircraft moving away from an observer
B the sound from an aircraft moving toward an observer
C the light from a galaxy moving towards an observer
D the light from an aircraft moving towards an observer

18 Which of these provides evidence to support the "Big Bang" theory?

A infrared radiation
B light
C microwaves
D ultraviolet radiation

19 The Universe is known to be:

A contracting
B expanding
C reproducing
D in a steady state

20 Two atoms have the same number of neutrons but different numbers of protons. They are:

A isotopes of the same element
B isotopes of different elements
C atoms of the same element
D atoms of different elements

21 The atomic number of gamma radiation is:

A -1
B 0
C $+1$
D $+4$

22 Radioactivity arises from the breakdown of an unstable:

A atom
B electron
C molecule
D nucleus

23 Which type of wave is most likely to be the cause of skin cancer?

A sound
B ultrasound
C light
D ultraviolet

24 An alpha-emitter is used in smoke alarms because it

A does not ionise air particles
B ionises air particles
C oxidises air particles
D reduces air particles

Answers to these questions can be found on pages 143–147

Getting it right

(a) The diagram shows an ultrasound scan of a fetus.

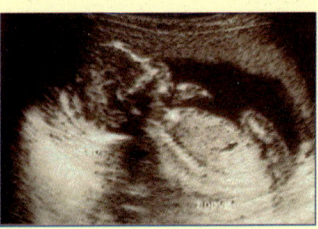

(i) Why is ultrasound preferred to X-rays for viewing a fetus?

Ultrasound is safe to use because it does not damage cells and tissues, but X-rays do. **[2]**

(ii) How is an ultrasound scan produced?

Pulses of ultrasound are emitted by a probe. The ultrasound is reflected by the fetus and these reflections are detected by the probe. A computer uses the information from the probe to build up a picture. **[3]**

Remember, this question is about the science of an ultrasound scan, so answers such as "the patient lies on a table...." would not be awarded any marks.

(b) The diagram compares the transmission of sound and electromagnetic waves under water.

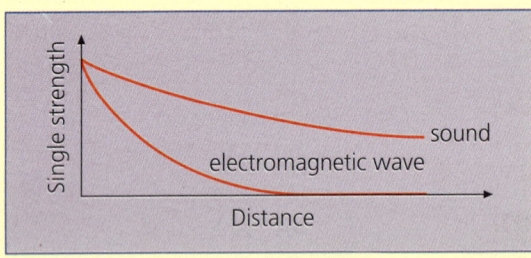

(i) Explain why ships use sound rather than radio waves to detect submarines.

Radio waves are electromagnet; these waves have a very short range in water. Sound waves can be detected as much greater distances. **[2]**

In open-ended questions like this one, it is important that you give as much detail as possible. You cannot lose marks by writing too much detail, but you can lose marks if you write too little.

(ii) Sound travels at a speed of 1500 m/s in water. A ship detects a submarine at a distance of 3000 m. How certain can the captain of the ship be about the position of the submarine?

The captain cannot be certain of the submarine's position. It takes 2 s for the sound that is reflected to travel back to the ship. In this time the submarine could have moved a considerable distance. **[3]**

Food production and the environment

All life depends on a supply of energy. Energy must enter and flow through an ecosystem for a range of organisms to survive.

The first topic in this module is about a key group of organisms that allow energy input into an ecosystem – the green plants.

Energy passes through an ecosystem via food chains and webs. This is the subject of the second topic. The diversity of organisms and their relationships in different ecosystems is outlined in the third topic.

Living organisms produce waste and finally die. The fourth topic explains the role of microorganisms in recycling.

Humans live throughout the ecosystems of the world and need nutrients to survive. The fifth topic is about how we achieve the maximum production of human food.

Control of plant activity

You need to know •••••••••••••••••••••••••••••••

✔ how plant cells differ from animal cells;
✔ an outline of photosynthesis and the key factors that affect the process;
✔ about the distribution and uses of sugars in plants;
✔ about the absorption and transport of water and minerals in plants;

✔ the role of guard cells in controlling water loss through transpiration;
✔ the role of plant hormones in controlling plant growth;
✔ a range of commercial uses of plant hormones.

How do plant and animals cells differ ••

These diagrams show typical plant and cells.

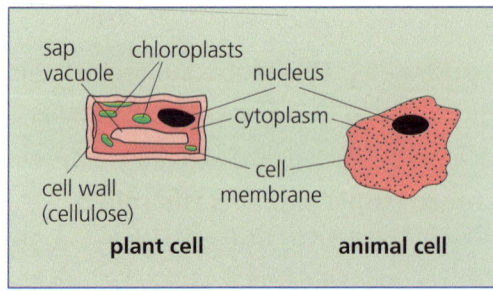

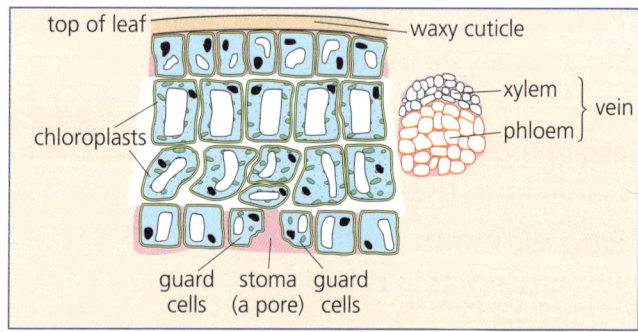

Cells with **chloroplasts** are found in various parts of plant shoots but there are more in the leaves. The section through a leaf in the diagram above shows three types of cell with chloroplasts.

There are more chloroplasts in the upper part of leaves, to absorb maximum light. The chloroplasts contain **chlorophyll**, a vital substance in the process of **photosynthesis**. Water for the process is stored in the vacuoles. The veins form a tubular transport system.

Learn the word equation for photosynthesis – you may be asked to complete it in an exam. Do **not** write chlorophyll along the equation line. Keep to the ingredients and products shown.

Photosynthesis •••••

Green plants have the ability to harness light energy using photosynthesis.

Photosynthesis produces

■ glucose, which contains energy (trapped as chemical energy)

■ oxygen as a waste product.

Limiting factors that affect the rate of photosynthesis:

■ the amount of light;

■ the amount of carbon dioxide;

■ the temperature.

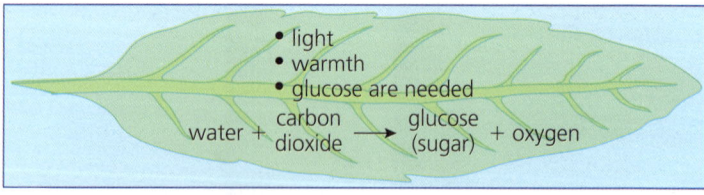

The process of photosynthesis

Plants have a transport system to supply glucose to other cells that do not have chloroplasts, e.g. root cells. In veins there is a special tissue, the phloem, which transports the glucose.
The tube-like cells in phloem transport glucose:
● in solution;
● to tissues that cannot photosynthesise;
● to energy storage areas, e.g. roots in carrots;
● to flowers and fruits so that the species can continue.

Plants need mineral salts

The roots absorb mineral salts needed for healthy growth.

Mineral	Used to make	Affects
nitrate	protein	growth
magnesium ions	chlorophyll	photosynthesis

How do plants take in water?

Roots absorb water through root hairs – cells with special adaptations to absorb water as well as minerals. Each root hair is a long thin cell that projects into the soil. Water is taken into a plant by the process of **osmosis**.

Diffusion

When a spoonful of sugar is put into a cup of tea the sugar molecules move so that they are in equal concentration everywhere, even without stirring. Molecules in fluids and gases move from an area of high concentration to where they are in low concentration. This is **diffusion**.

Osmosis

This is a special type of diffusion that takes place when a solution of higher concentration is separated from a solution of lower concentration by a **selectively permeable membrane**. This has tiny holes that **only** allow water molecules to pass through. The water molecules move from the less concentrated solution to the more concentrated solution. The process depends on the concentration of solutions inside and outside the cells.

Water transport in plants

Once inside a plant, water is transported to all cells. Across a root, water goes from cell to cell by osmosis. Water reaches the middle of a root and enters tube-like vessels known as the **xylem**. These vessels consist of dead hollow cells. Each cell is joined to the next to form a long tube, ideal for water transport.

A plant needs water:
- for photosynthesis, where it is used with carbon dioxide to make glucose;
- for all enzyme-based reactions;
- to give its cells hydrostatic strength, known as **turgidity**.

Plants lose water by the process of **transpiration**. Water vapour diffuses through pores in the leaf, called **stomata**. Water evaporates from many surfaces of a plant, but much more is lost by the stomata. Open stomata allow more water loss. Stomata can open and close in response to different conditions. Most plants close their stomata at night.

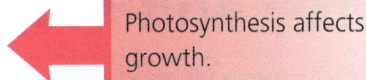

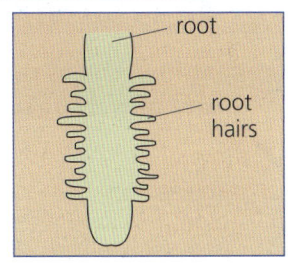
Photosynthesis affects growth.

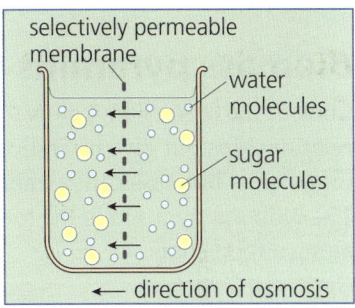
The process of osmosis

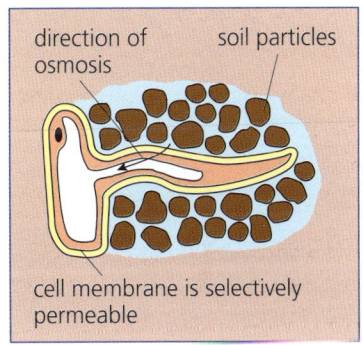
Absorption of water by a root hair – water moves into the root hair until the concentration of water molecules in the sap vacuole is equal to the concentration of water

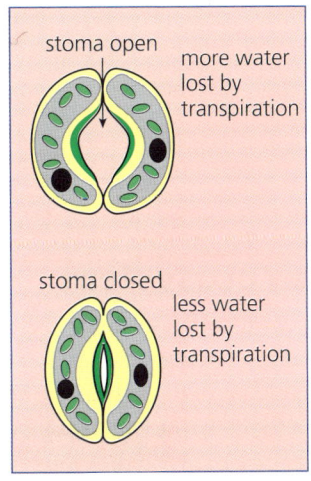

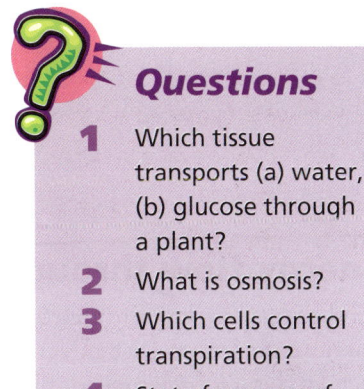

Questions

1 Which tissue transports (a) water, (b) glucose through a plant?

2 What is osmosis?

3 Which cells control transpiration?

4 State four uses of glucose in a plant.

Energy and ecosystems

You need to know •

✔ that food chains can be represented quantitatively by biomass pyramids;

✔ how energy is passed through an ecosystem;

✔ how energy is lost along a food chain;

✔ that mammals have problems with heat loss.

Biomass pyramids • • • • • • • • • • • • • • • • •

Look at the food chain below. The arrows show the feeding relationships but give no indication of numbers or size of the organisms. Biomass can be measured and is the mass of a species. The biomass pyramid is the total mass of each species at each level along a food chain.

> *Remember to put the arrows in the right direction in exam questions.*

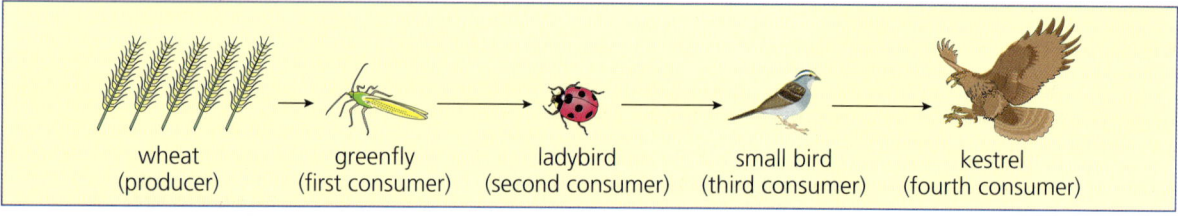

| wheat (producer) | greenfly (first consumer) | ladybird (second consumer) | small bird (third consumer) | kestrel (fourth consumer) |

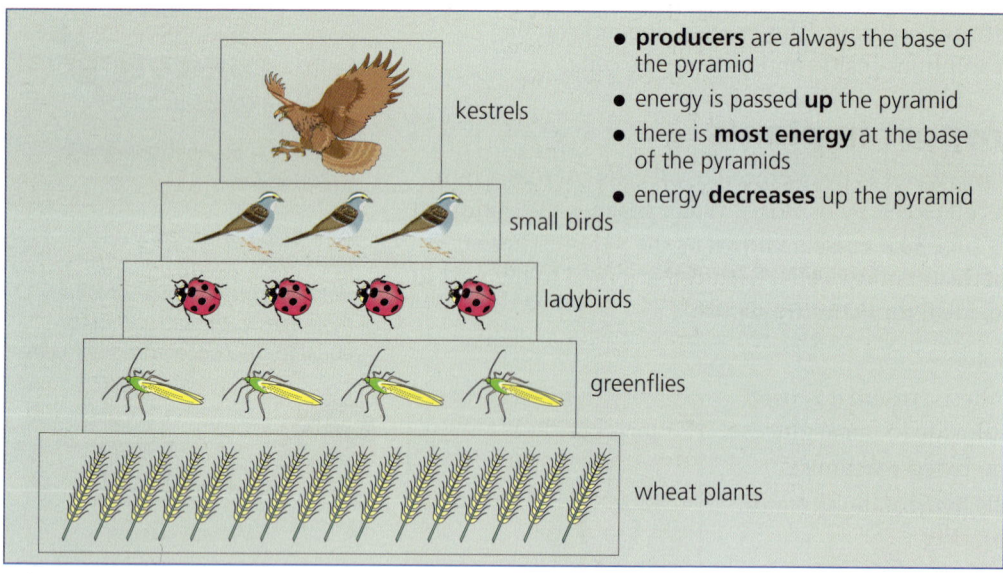

- **producers** are always the base of the pyramid
- energy is passed **up** the pyramid
- there is **most energy** at the base of the pyramids
- energy **decreases** up the pyramid

kestrels

small birds

ladybirds

greenflies

wheat plants

A biomass pyramid

Energy flow through an ecosystem • • • •

Food chains interlink through food webs in an ecosystem. Energy enters an ecosystem via producers. Each species at each level is a food source for the next consumer – right up to the top consumer. Energy is lost throughout a food chain. The next level of organisms always has less available energy than the previous one.

Warm-blooded animals need a lot of energy to maintain their body heat. Mammals lose a lot of heat energy to the air.

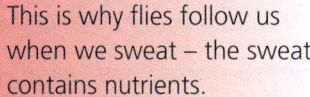

Energy is lost along a food chain because:

■ respiration releases heat energy so it is not available for the next level;

■ organisms excrete substances, e.g. sweat and urine, which loses energy.

> This is why flies follow us when we sweat – the sweat contains nutrients.

Natural ecosystems

An ecosystem consists of:

■ an area that has particular characteristics, such as desert;

■ a number of organisms living together in a community;

■ the abiotic factors of the environment, such as the water content of the soil;

■ a stable environment where substances are recycled, helping to maintain the community.

Biodiversity

Even a desert ecosystem supports a wide range of different living things. This is known as **biodiversity**. Rainforest has even greater biodiversity because of the greater energy input into this ecosystem. Each type of organism contributes to the success of the ecosystem. Producers allow the energy flow in and decomposers recycle nutrients from waste products and dead organisms.

> Biodiversity means different species in an area.

The more plant species there are in an area, the more they supply different habitats and different nutrients. This results in greater biodiversity. In areas where only one plant species is grown then biodiversity is decreased, e.g. commercial woodlands in the UK. Scot's pine trees are grown in huge forests. They support a very limited number of other species. Mixed woodland supports a wide range of different species.

Every species has needs – these must be met if a species is to survive in an area. For example, rabbits need plants to feed on, soil to build underground warrens, oxygen to breathe and a suitable temperature.

The adaptations of organisms to their environment determine their relative numbers. The better adapted they are, the more successful they are. The more food there is, the better the conditions are, and the more successful they are.

> You may be given different habitats to analyse in your exams. The principles will be the same. One plant species in an area only supports a narrow range of consumers.

Questions

1 Which organisms are at the base of the pyramid of biomass?

2 How is energy lost along the food chain?

3 Explain the effect monoculture has on biodiversity.

Action of microorganisms in ecosystems

You need to know ●

✔ about the process of decay;

✔ the carbon and nitrogen cycles;

✔ the effects of deforestation;

✔ the importance of conservation.

Decomposers and the compost heap ●

Material such as leaves and animal faeces are put in a heap. Decomposing bacteria and fungi begin to rot down the material by secreting enzymes that break down the organic material. They breed in large numbers in the heap. The heap gets smaller and is full of minerals, so is useful as a fertiliser for plants.

Some garden centres sell packets of bacteria as a starter for compost heaps. However, they are not really necessary – bacteria are in the air, water and soil all around us.

After several days, decomposers such as bacteria become established and increase in numbers as they reproduce. Decay speeds up and the heap may steam. This is caused by the heat energy released during respiration of the microorganisms.

Requirements for decay	How does this help decay?
presence of decomposer bacteria and fungi	if there were no microorganisms then there would not be any decay!
oxygen	many decomposers respire aerobically
moisture	needed if the enzymes produced by the decomposers are to work
suitable temperature	needed if the enzymes produced by the decomposers are to work

The carbon cycle ● ● ● ● ● ● ● ● ● ● ● ● ● ● ● ● ● ●

The air contains about 0.03% carbon dioxide. The amount is kept in balance by the processes involved in the carbon cycle.

There are three key processes in the carbon cycle:
■ Photosynthesis, which removes carbon dioxide from the air, fixing it in carbohydrate molecules.
■ Respiration, which breaks down organic compounds, releasing carbon dioxide back into the air.
■ Combustion, which also releases the carbon dioxide back into the air.

The carbon cycle

Deforestation ● ● ● ● ● ● ● ● ● ● ● ● ●

The human population of the world continues to increase. This means that we need more areas for houses, factories and farming and has led to deforestation – the cutting down of large areas of trees. There are several disadvantages to deforestation, including:

- it reduces the amount of oxygen produced by photosynthesis;
- it destroys the habitats and food sources of many organisms;
- tree roots are no longer able to bind the soil, leading to soil erosion and land slip;
- it increases levels of carbon dioxide in the air, leading to global warming.

Too much burning and not enough photosynthesis? It is important that processes remain in balance. Cutting down the rainforests is not a good idea. Conservation means that we should keep these habitats.

The nitrogen cycle

The element nitrogen is needed in all amino acids and, therefore, proteins. This nitrogen is of no direct use to most organisms but it can be converted into a useful form during the nitrogen cycle.

Nitrogen-fixing bacteria – these bacteria live in nodules in the roots of plants in the pea and bean family. They use nitrogen from the air and help the plant make nitrogen compounds. As a result the plant goes on to make protein. Some plants may be eaten by animals.

Plants need other ions as well as nitrates, e.g. phosphates.

Decomposers – some bacteria and fungi are able to break down dead plants and animals as well as faeces and urine. They produce ammonium ions as a waste product.

Nitrifying bacteria – these use the ammonium ions. They produce nitrates, very useful as fertiliser for plants.

Denitrifying bacteria – these change nitrate ions and release nitrogen gas back into the atmosphere.

Nitrate fertilisers – these supply plants with the element nitrogen in a suitable form to make amino acids, then proteins. Fertilisers improve the yield of our crops.

The nitrogen cycle

In an exam you may be asked to complete an empty box in a nitrogen cycle diagram. Rarely will the full nitrogen cycle be given. Do not mix up the different processes.

Questions

1 What are the optimum conditions for decay?
2 Why is it necessary for materials to be decomposed?
3 Why is a compost heap turned over halfway through the decay process?
4 State the three main processes in the carbon cycle.
5 Give reasons for deforestation.
6 What is nitrogen fixation?

Maximising food production

You need to know ••••••••••••••••••••••••••••••••••••

✔ how we use a range of chemicals to achieve a high crop yield;

✔ about the dangers of overusing pesticides;

✔ how we use biological control in crop production.

Using chemicals to increase crop yields

Fertilisers provide plant nutrients important for healthy growth. A crop can also be given chemicals that can result in a maximum yield. The table below shows the effect of each chemical.

Type of chemical	What is its effect?	How does it help achieve a high yield?
pesticide	kills pests like greenfly and mice	reduces damage to crop by the pest
insecticide	kills insect pests like greenfly	reduces damage to crop by the insect pest
herbicide (weedkiller)	kills weeds	weeds compete for light, water and minerals, so with the weeds gone, there are more of these factors available for the crop
fungicide	kills fungi	some fungi attack the crop and feed on its cells – this damage is reduced by using a fungicide

Problems with agricultural chemicals

If farmers are to meet the food requirements of the population, they have to use chemicals. However, these chamicals can remain in the crops. When we eat an undamaged apple it is likely that it still contains a residue from the chemicals it was sprayed with.

Pesticide is a general term given to chemicals used to kill animals that damage crops. For example, slug pellets kill slugs and snails. Insecticides such as dimethoate can also be considered as a pesticide.

If chemicals are passed on to consumers they can have toxic effects on their bodies. Organophosphates have been linked to serious human diseases. Even cancers are linked to agricultural chemicals.

Some farmers prefer to grow crops without using any man-made chemicals, but they still use natural chemicals such as pyrethrum – a natural chemical from the plant species of the same name. This is known as **organic farming**.

Selective herbicides kill the weeds but not the crop.

Fungicides are often used as a seed dressing. They prevent seeds from being destroyed by fungi during germination.

Biological control

Another method can be used to get rid of pests that presents no dangers for the consumer – **biological control**. There are several variations of this method, including using a predator to kill a pest (prey). For example, predatory wasps are used to kill greenhouse whitefly.

Note that you never use the predator method of biological control and a pesticide. The predators may be killed by the chemical.

Crops such as tomatoes are attacked by a sap-sucking insect, the whitefly. Whitefly females lay eggs on tomato leaves. Larvae hatch from the eggs. A predatory female wasp lays an egg into a whitefly larva. Days later an adult wasp emerges from the larva. The wasp has used the larva as a food source and only its outer shell is left.

The wasps breed quickly – a big advantage. An insecticide usually needs several applications, but this method of control needs just one introduction at the beginning – nature does the rest.

Another method of biological control is the release of sterile males. This method relies on the fact that some insects mate only once. For example, the New World screw fly attacks farm animals (and people) in some countries. It lays eggs in cuts and the larvae attack the body internally. Eventually it can kill the victim. The answer has been to subject adult flies to a dose of radiation. This causes every male to become sterile. Released in large numbers into problem areas from aeroplanes, each sterile male mates with a local female. She lays sterile eggs that never hatch out, and the problem is reduced.

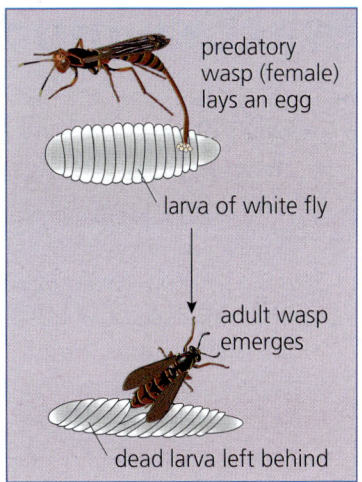

predatory wasp (female) lays an egg

larva of white fly

adult wasp emerges

dead larva left behind

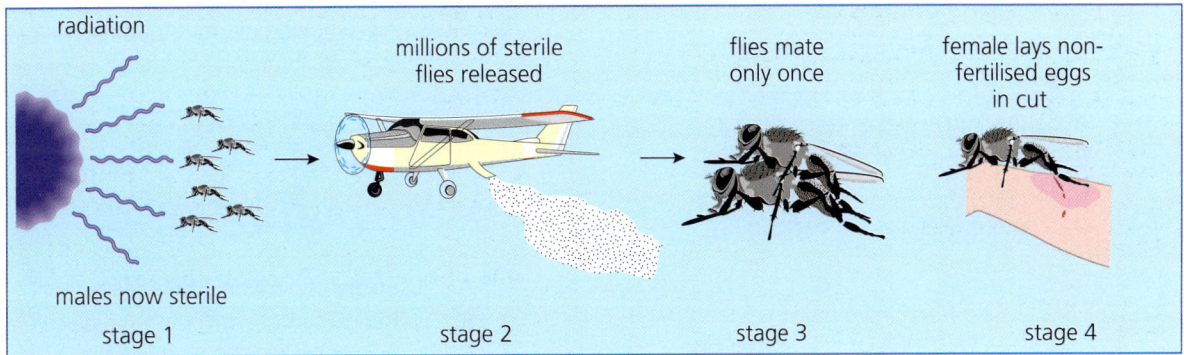

radiation

millions of sterile flies released

flies mate only once

female lays non-fertilised eggs in cut

males now sterile

| stage 1 | stage 2 | stage 3 | stage 4 |

Greenhouse management • • • • • • • • • • •

In a country where conditions are warm enough for plants, then greenhouses are not necessary. In the UK there are cold months where frosts can kill many plants, so greenhouses are used as a controlled environment.

Growing plants in greenhouses is known as **intensive farming**. The aim is to provide optimum conditions for the growing crop. Large commercial greenhouses are used to supply supermarkets. Large amounts of the crop can be transported at the same time, which reduces costs. It is important in intensive crop production that costs must be justified. Profits need to be maximised.

Questions

1 A slective herbicide kills weeds but not the crop. What is the advantage of this?

2 What is the advantage of organic farming?

3 Why should pesticides *and* biological control *not* be used together?

4 Give one way that a greenhouse can increase the yield of a crop.

Practice module test

You will have 17 minutes to answer these questions

1 The diagram below shows a Visking tubing bag used to demonstrate osmosis. The bag containing sugar solution of higher concentration was placed in another sugar solution of lower concentration. The bag increased in mass because:

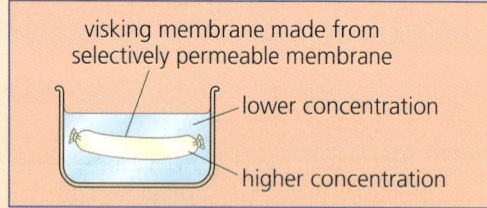

visking membrane made from selectively permeable membrane

lower concentration

higher concentration

   A sugar molecules entered the bag
   B water molecules entered the bag
   C sugar solution entered the bag
   D water molecules left the bag

2 Pots X and Y each contained a wheat plant and a weed. Pot X was sprayed with a selective herbicide. Pot Y was not. The diagrams show the plants before spraying and a week after spraying. The wheat grew very well because:

wheat weed

pot X   pot Y

before   after   before   after

   A the weed was killed, providing more minerals for the wheat
   B the selective herbicide supplied more nutrients to the wheat
   C the wheat received more water than the weed
   D the wheat received more light than the weed

3 Vaseline makes a surface waterproof. Equal masses of leaves of the same plant species were treated with Vaseline in different ways. They were left in the same conditions.

Group 1 – Vaseline on upper surface

Group 2 – Vaseline on lower surface

Group 3 – Vaseline on upper and lower surfaces
The leaves were weighed at the start and after 2 weeks. Beginning with the greatest first, the correct order of mass would be:

   A 1 2 3
   B 3 2 1
   C 3 1 2
   D 2 1 3

4 The diagram below shows a test tube containing some water and pondweed. The tube was given a supply of light. Which statement correctly describes what is taking place in the tube?

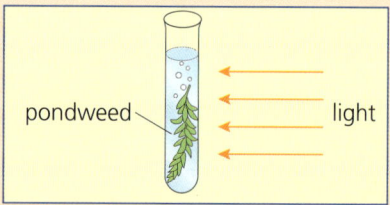

pondweed   light

   A just respiration would take place
   B just photosynthesis would take place
   C both respiration and photosynthesis take place
   D transpiration and photosynthesis are taking place

5 Which of the following gives one loss of energy from a food chain?
rose bush → greenfly → ladybird

   A photosynthesis
   B respiration
   C greenfly are eaten by a ladybird
   D a rose bush is attacked by greenfly

6 Which one of the following is reduced as a result of deforestation?

   A oxygen content in the air
   B carbon dioxide in the air
   C nitrogen content of the air
   D the chances of soil erosion

7 Biomass of producers is increased by:

   A photosynthesis
   B respiration
   C transpiration
   D excretion

8 The diagram shows a plant cell with maximum water content. Which of the following statements is not true?

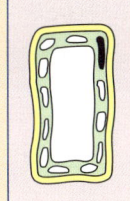

  A the cell is turgid
  B the cell has a maximum strength
  C the cell has minimum strength
  D the sap vacuole has maximum water content

9 Sugar is made in a leaf. It is transported to the roots through:

  A xylem
  B epidermis cells
  C phloem
  D palisade cells

10 The diagram below shows the underside of a leaf, seen under a microscope. The number of guard cells shown is:

  A 30
  B 15
  C 28
  D 14

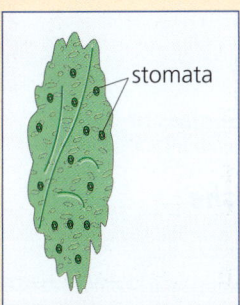

stomata

11 Ann investigated the rate of photosynthesis in pondweed at different temperatures. Carbon dioxide was bubbled through the water throughout the investigation. The results are shown on the graph below. What limits the rate of photosynthesis at position X?

  A oxygen
  B carbon dioxide
  C water
  D temperature

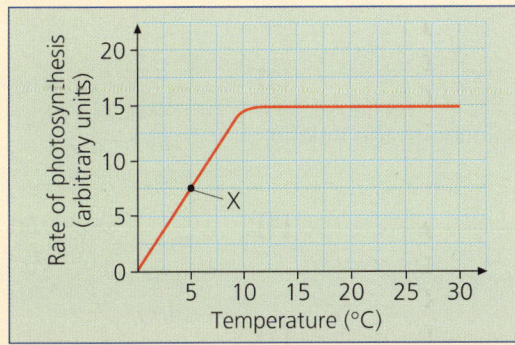

12 The sequence below shows part of the nitrogen cycle where nitrification takes place. In a compost heap **all** of the Nitrosomonas bacteria die. Which of the following statements is **not** true?

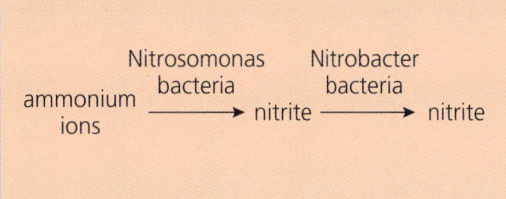

ammonium ions → Nitrosomonas bacteria → nitrite → Nitrobacter bacteria → nitrite

  A no more nitrate is produced
  B only a small amount of nitrite is produced
  C ammonium ions build up
  D Nitrobacter increase in numbers because they have more to feed on

13 The rate of transpiration changes with temperature and humidity of the atmosphere. In which of the conditions below would the rate of transpiration be lowest?

	Temperature of air (°C)	Humidity of atmosphere (% saturation)
A	30	75
B	40	25
C	0	80
D	45	50

14 A lily plant is cloned using tissue culture. A small piece of bud is put into a tube with a nutrient medium. After a few weeks the cells reproduce to form a callus. What happens during the next stage of development?

  A the callus is transferred to another tube
  B the callus is carefully put into a plant pot with compost
  C roots and leaves begin to grow, forming a tiny plant called an explant
  D more nutrient medium is added

Answers to these questions can be found on pages 143–147

Getting it right

1 Kylie investigated a process using Visking tubing. She used the tubing to make bags. For each bag, she

■ knotted one end of the tubing

■ put in the same volume of sugar solution

■ knotted the other end of the tubing to complete the bag

■ weighed each bag

■ put five bags into water at each of the following temperatures: 20°C, 30°C, 40°C

■ left each bag for 1 hour, then re-weighed it

Each bag increased in mass. The graph shows the effect of different temperatures on the average increase in mass of the bags.

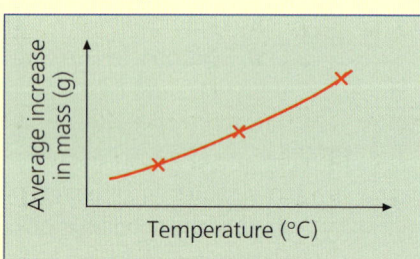

Every point plotted on this graph is an average of five measurements. These are more likely to be true representations of the effect of temperature.

(a) What effect did temperature have on the increase in the mass of the Visking tubing bags?

The greater the temperature, the greater the increase in mass. [1]

(b) Explain why each bag increased in mass.

Water molecules entered each bag; bag is selectively permeable osmosis takes place. [3]

Always read the details carefully. In this question you are given a lot of information and have to make a connection in your mind. Osmosis is the key. Always ask yourself "Which part of the specification is this question from?"

(c) Why did Kylie test five bags at each temperature?

She found the average so that each average increase accurately showed that true value; less chance of fluke values not being a true representation. [3]

(d) (i) What could Kylie have used to measure the volume of sugar solution?

A measuring cylinder. [1]

(ii) How could she have transferred the sugar solution to the Visking tubing?

A dropping pipette. [1]

Explain means exactly that – give three details here to score your three marks. Always look at the mark value.

(e) What could Kylie have used to maintain each temperature?

An electric water bath. [1]

Health and exercise

8

Our lifestyle can either enhance or harm our health. This module covers some of the body's major systems and associated health issues.

The first topic in the module is about the structure and function of the respiratory system and the problems associated with smoking tobacco.

Chemicals such as nutrients and hormones are transported around the body. The second topic covers the structure and functions of the circulatory system and associated health problems.

Energy released inside our cells gives us life. The third topic explains how respiration releases energy and how we can maintain our health by exercising.

Finally, two important health problems are explained – drug misuse and inherited disease.

The lungs

The human respiratory system • • • • • • • •

The complex structures of the human respiratory system have two functions:

■ to supply the blood with oxygen;

■ to remove toxic carbon dioxide from the blood.

The diagram below shows the human **thorax**, the name for the chest area. The continuous action of breathing in and out is known as **ventilation**.

The table below explains the structure and function of the various parts of the human respiratory system.

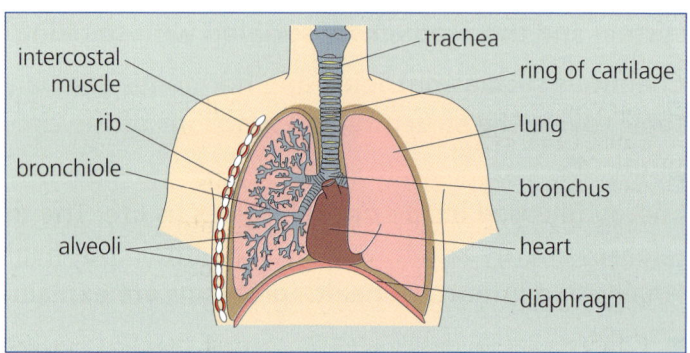

The human thorax

Structure	Function
trachea	takes air into and out of the lungs
cartilage ring	supports the trachea, keeping it open and preventing it collapsing
bronchus	transports air, connects trachea to bronchioles
bronchioles	tubes which connect bronchi to alveoli
alveoli	one cell thick, large surface area, covered by capillaries, oxygen diffuses into blood and carbon dioxide out of blood
diaphragm	a band of muscle that can make air pressure in the lungs lower or higher than that outside body
ribs	protect lungs and heart, help to lift or lower chest
intercostal muscles	can lift chest upwards and outwards or downwards and inwards
lungs	key organs of gaseous exchange

As we breathe in:

■ the diaphragm contracts and flattens;

■ the intercostal muscles contract, causing the rib cage to move upwards and outwards;

■ this creates a lower pressure in the lungs than in the air outside, so air flows in through the nose/mouth, down the trachea into the lungs.

Learn this sequence carefully – it is likely to be worth 6 marks in an exam!

Clean air

The environment inside the nose cavity is moist and warm. Goblet cells in the membranes produce a fluid – **mucus**. The mucus traps dust particles and bacteria. The cilia move the mucus to the back of the throat, where the "dirty" mucus is swallowed, or coughed up.

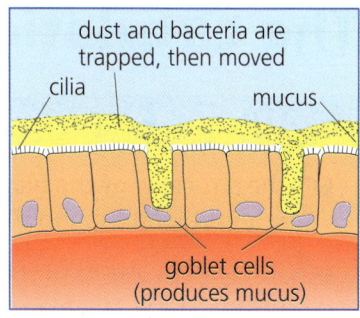

dust and bacteria are trapped, then moved

cilia

mucus

goblet cells (produces mucus)

Smoking can seriously damage your health

Smoking tobacco – cigarettes in particular – is dangerous. There are so many risks associated with tobacco that health deterioration is a certainty.

Smoking causes serious diseases.

Do not miss out the mucus – it is needed to help the cilia move the dust.

Cancer

- Tar released by burning cigarettes contains substances that cause cancer (**carcinogens**).

- Carcinogens can affect the DNA in cells.

- Normal cells in the lungs can change into cancer cells.

- Cancer cells grow quickly and are abnormal, preventing many organs from working properly.

- Cancer in the lungs usually occurs first, but new cancers can appear around the body as **secondary cancers**.

Cancer cells may be destroyed by chemotherapy or radiation treatment, but many forms of cancer are incurable.

Emphysema

- The alveoli walls become less elastic.

- This results in a much smaller surface area for the exchange of oxygen and carbon dioxide.

- Taking less oxygen into the blood limits the amount of energy released in the body.

- The emphysema sufferer is much less active and really struggles to breathe in and out.

- Symptom relief is by breathing in pure oxygen, usually from an oxygen cylinder with face mask.

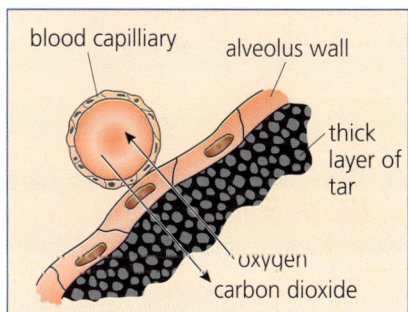

blood capilliary

alveolus wall

thick layer of tar

oxygen

carbon dioxide

An emphysema sufferer may not be able to blow out a lighted match, even when it is straight in front of their mouth.

Fighting the addiction

The nicotine in cigarettes is highly addictive. Giving up needs determination. To support someone trying to give up there are nicotine patches and nicotine gum. They provide tiny doses of nicotine that help some smokers to gradually stop the habit.

Questions

1 What does the diaphragm do during breathing in?

2 What is the function of the alveoli?

3 What are the advantages of breathing in through the nose rather than through the mouth?

The circulation

You need to know ••••••••••••••••••••••••••••••

✔ the structure of the heart and how it pumps blood around the body;

✔ the main features of the circulatory system;

✔ how red blood cells transport oxygen;

✔ the structure and function of arteries and veins;

✔ about the circulation problems associated with our lifestyle;

✔ how the hormone insulin regulates glucose.

The human heart – a powerful pump •••

The heart moves blood around the body. It consists of **cardiac muscle**, which never tires, throughout your lifetime. The heart function sequence is:

■ the heart receives deoxygenated blood via the veins;

■ it pumps blood to the lungs;

■ the blood picks up oxygen at the lungs and carbon dioxide is given up;

■ the heart receives oxygenated blood from the lungs;

■ it pumps blood all around the body.

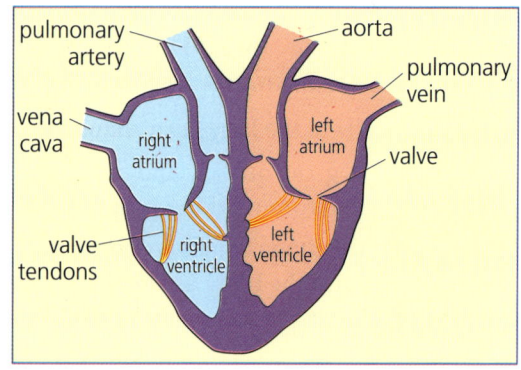

The human heart

The heart valves

■ The function of all valves is to make sure that blood moves in **one direction only**. Any attempt by the blood to move backwards and the valve closes tightly.

■ The **valve tendons** prevent each atrio-ventricular valve from collapsing back into an atrium.

■ **Semi-lunar valves** are in the aorta and pulmonary artery, and prevent back-flow of blood into the atria.

The coronary artery supplies the heart with its own supply of oxygen and glucose.

How our lifestyle affects our health

Our diet and behaviour affect our health considerably.

The consumption of **animal fat** (**saturated fat**) can lead to damage to the blood vessels (**arteriosclerosis**):

■ yellow fat streaks build up under the lining cells of a vessel;

■ an **atheroma** develops that is built up from cholesterol absorbed from blood;

■ the atheroma develops into a white fibrous lump;

■ eventually, it raises the blood pressure and can cause a blockage.

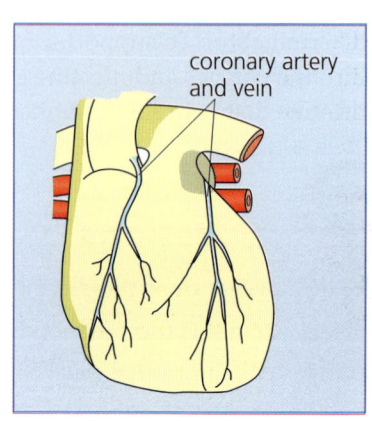

Heart disease – a possible cause

- A **thrombosis** (blood clot) may block a coronary vessel, preventing supplies such as oxygen reaching the heart.

- If part of the heart is deprived of oxygen, it causes a sharp pain (**angina**).

- This pain is a warning to seek medical help.

- If not treated, a total blockage may occur, resulting in a **heart attack**, which can be fatal.

1 ⎨ — a normal, healthy artery

2 ⎨ — atheroma (a fatty lump)

clot

3 ⎨ — atheroma
this artery is almost blocked

Remember that smoking also damages your blood vessels. They become less elastic and can sometimes leak, causing internal bleeding (haemorrhage).

Blood sugar (glucose) – the full story • • •

We may eat a chocolate bar, full of glucose, or a slice of bread high in starch. Both foods result in the absorption of a high level of glucose into the blood. The key to regulation of glucose is an important organ, the **pancreas**. In response to a high amount of glucose in the blood it produces a hormone, **insulin**. This is secreted into the bloodstream.

Insulin has the following effects:

- it reduces the amount of glucose in the blood
- it allows glucose into cells where it is respired to release energy;
- it allows excess glucose into the liver;
- in the liver glucose is changed into glycogen;
- glycogen is an energy store;
- glycogen can later be changed back to glucose, when the person has a low level of glucose in the blood.

Changing the glycogen back to glucose is controlled by a different hormone.

If we do not produce enough insulin then glucose in the blood builds up to a dangerous level. This condition is known as **diabetes**.

The symptoms of diabetes include glucose in the urine, excessive urine production and great thirst. It can eventually lead to blindness.

A diabetic person can still regulate blood glucose, by

- adjusting the diet to limit the amount of carbohydrate consumed;
- injecting insulin into the bloodstream every day.

Doctors prescribe the amount of insulin that the diabetic is lacking. Injections are needed every day, for the rest of the person's life.

Insulin is a protein and so cannot be taken by mouth – it would be digested.

In the early days, when insulin was prescribed it was obtained from the pig pancreas. Today, human insulin is used, as a result of genetic engineering.

Questions

1. What effect does insulin have on glucose in the blood?
2. Why are polyunsaturated fats recommended to avoid circulation problems?
3. What is angina?

Respiration, energy and exercise

Respiration •

Glucose is a source of chemical energy. Energy must be released for cell activities. Respiration releases this energy. There are two types of respiration – aerobic respiration and anaerobic respiration.

Aerobic respiration

Oxygen must reach the cells if this process is to take place. Oxygen diffuses from the red blood cells through capillary walls into cells. Aerobic respiration takes place in cytoplasm plus the mitochondria.

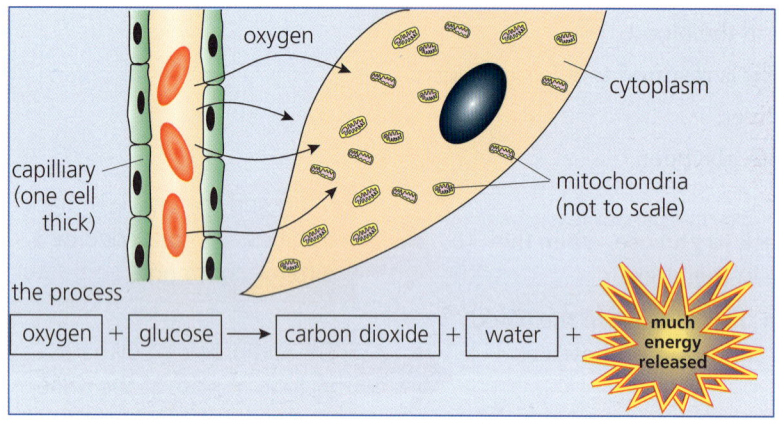

The process of respiration

How do heart rate and breathing rate affect aerobic respiration?

When exercising or under stress, adrenaline is produced by the adrenal glands. This increases both the heart rate and the breathing rate. As a result, beginning at the lungs:

■ more oxygen diffuses into the blood;

■ oxygen is transported to the muscles more quickly;

■ the rate of respiration increases as more oxygen and more glucose reach the cells.

At the same time, carbon dioxide must be excreted faster:

■ more carbon dioxide diffuses into the blood from the cells;

■ carbon dioxide is transported to the lungs more quickly;

■ more carbon dioxide diffuses into alveoli of the lungs to be lost during breathing out.

Try to remember this word equation. It is the form of respiration that releases most energy.

Think about this logically – as you breathe in and out more often you are bound to take in more oxygen and give out more carbon dioxide.

Anaerobic respiration

Imagine a 100 metre sprint race. A sprint athlete needs maximum energy release at the muscles. Energy demand for a race like this, over in a few seconds, cannot be met just by aerobic respiration. You cannot take in a large enough amount of oxygen in just a few seconds. Anaerobic respiration releases this extra energy requirement.

glucose → lactic acid + small amount of energy released

Lactic acid quickly builds up in the muscle tissues and some reaches the blood. Build-up of lactic acid can give a pain, known as cramp. The legs become more difficult to move and you meet a pain barrier. It is lactic acid build-up around the muscles that causes this. Most people slow down as a result.

At the end of the race, repayment of oxygen debt takes place. Rapid breathing rate at the end of the race results in a greater supply of oxygen. Most of the lactic acid is oxidised to carbon dioxide and water, releasing energy.

During repayment of oxygen debt, lactic acid is involved in different processes:

Process	Percentage of lactic acid
oxidised to carbon dioxide and water	64
changed to protein	11
changed to glucose	6
changed to glycogen in muscles, liver	19

The following table provides a comparison of aerobic and anaerobic respiration.

	Aerobic respiration	Anaerobic respiration
substrate used	glucose	glucose
energy released	bigger amount	lower amount
waste product(s)	carbon dioxide + water	lactic acid

In a long distance race, more aerobic respiration is used.

Many students forget the details of respiration. They remember oxygen but not the carbon dioxide – do not make this mistake.

The legs feel heavier! Trained athletes build more capillaries in their muscles. More oxygen reaches the lactic acid. Less pain results. Cramp is much less likely.

You will often be asked about the differences between the two processes in an exam.

Misuse of steroid drugs ● ● ● ● ● ● ● ● ● ● ●

Some young athletes take steroid drugs. They promote muscle growth and improve strength. However, there are major problems with the use of these hormones – not only are they illegal, but they have long-term side-effects. They can increase the risk of both heart disease and sterility.

Questions

1 Complete this equation for aerobic respiration
_____ + _____ → carbon dioxide + water

2 What effect would an increase in heat rate have on aerobic respiration?

3 Which substance is produced in the muscles due to anaerobic respiration?

Inherited diseases

You need to know •

✔ the symptoms of a range of inherited disorders;

✔ how gene therapy can be used to treat some genetic disorders;

✔ how genetic counselling can help people make informed decisions.

Harmful microorganisms cause many common diseases, but some are caused by alleles that do not function properly. They are passed on from parents to their children.

Cystic fibrosis is an inherited disease carried by a recessive allele. Its characteristics are:

- a thick sticky mucus is produced in the lungs;
- breathing is difficult;
- the sticky mucus must be coughed up each day to relieve breathing difficulty;
- fluid that lubricates the first part of the small intestine is also thick and sticky;
- food passes through the intestines with greater difficulty and digestion is inhibited.

Remember a gene is a section of DNA that has a specific function. Each gene has a number of different forms, e.g. "normal mucus" and "cystic mucus".

We inherit a genetic disease when a gene does not function correctly. Every nucleus in our body cells has the gene. **Gene therapy** is the attempt to replace defective genes with copies of the healthy ones. Gene therapy is in the early days of development. In the case of cystic fibrosis, healthy genes are sprayed into the lungs via an aerosol. Surface cells receive the healthy gene and no longer produce the sticky mucus. Unfortunately, this treatment is only effective for about a month. It is still experimental.

Replacement of every gene that does not function correctly is an extremely difficult task and not yet achieved.

Sickle cell anaemia is another inherited disease carried by a recessive allele. In people who suffer from the condition, normal haemoglobin (dominant) is replaced by a different type, haemoglobin-S. The characteristics of sickle cell anaemia are:

- red blood cells distort and become sickle-shaped;
- these cells are less efficient than normal red blood cells at transporting oxygen;
- as a result, the patient suffers from anaemia;
- the sickle red cells tend to "log-jam" and block capillaries, depriving cells of oxygen;
- some people have the normal allele and the sickle cell allele – this is known as sickle cell trait, because some of the red blood cells are normal and some are sickle-shaped (30–40%), but the condition is not life-threatening;
- some people inherit sickle alleles from both parents – this is often fatal.

People who have both alleles who are cystic often die.

Inheritance of a genetic disease

If the disorder is controlled by a **dominant allele** then it will be much more common in the population. If the disorder is passed on by a **recessive allele** then the disease will appear less often in the population.

People with sickle cell disease often die. People with sickle cell trait have symptoms. They cannot transport oxygen as well, but the trait condition is not life threatening. Analysis of family gene profiles can help potential parents to make suitable decisions.

Genetic counselling

Doctors can analyse the alleles across families. This is **pedigree analysis**. In counselling sessions genetics experts can inform potential parents of the chances of their children being carriers or having a disease.

Two people who have cystic fibrosis have a 100% chance of producing a cystic child. They have a difficult decision to make. Equally, two people who are known carriers have a 1 in 4 chance of producing a child who suffers from the condition.

Moral dilemmas

Family genetics problems need to be handled sensitively. Couples need to make up their own minds about having children. There is a dilemma. If decisions are made not to have children, knowing they have a defective gene, then the number of defective genes in the population decreases. There are religious and cultural factors that can also be part of the decision-making process.

In recent years there has been a lot of progress made in research into human genes. In particular, the Human Genome Project has identified all of the human gene positions along our chromosomes.

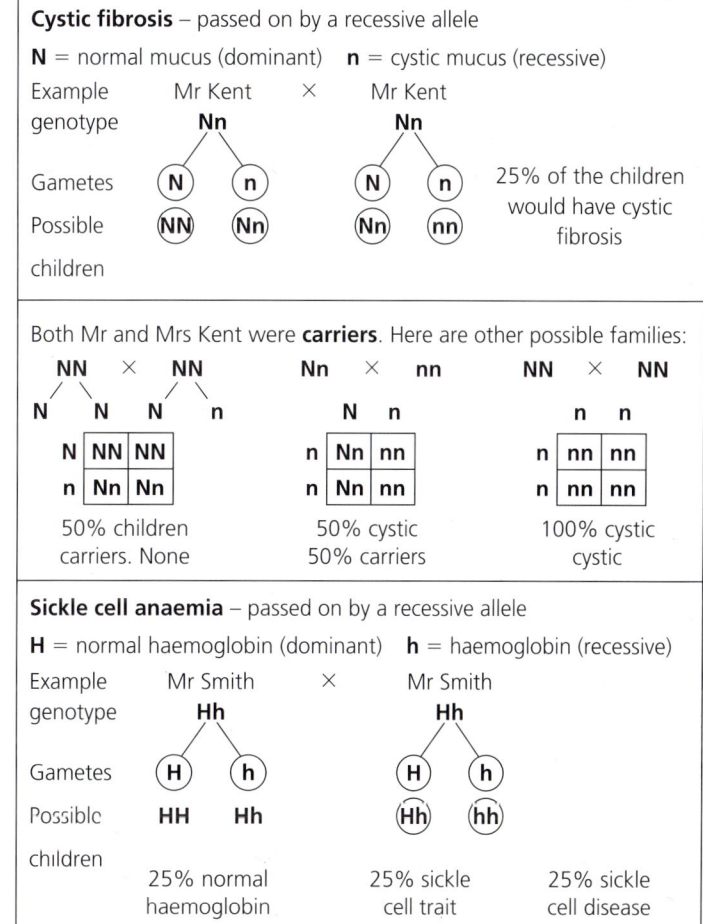

Some genetic research is controversial. It makes use of human fetuses. Some people disagree with their use, but there are potential advantages. If we find cures for major disease then millions of people in the future may benefit.

Laws have been passed to licence research using fetuses under a strict ethical code. Even so the spiritual and moral dilemma continues.

Questions

1 Name one inherited disease which is controlled by a dominant allele.

2 What are the chances of two carriers having a child with cystic fibrosis?

3 What is gene therapy?

Practice module test

You will have 17 minutes
to answer these questions

1. Oxygen passes from the air in the lungs into the
   blood by:
   - A   osmosis
   - B   diffusion
   - C   respiration
   - D   breathing in

2. Look at the diagram
   of the heart below.
   Which vessel takes
   deoxygenated blood
   into the heart?
   - A   vena cava
   - B   aorta
   - C   pulmonary artery
   - D   pulmonary vein

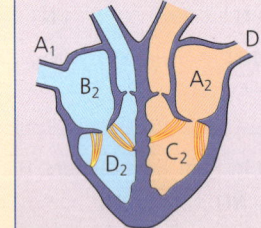

3. Look at the diagram of the heart above. Which
   chamber contracts to pump oxygenated blood
   to most parts of the body?
   - A   left atrium
   - B   right atrium
   - C   left ventricle
   - D   right ventricle

4. The diagram below shows some of the cells
   found in the lining of the lungs. The cilia move
   backwards and forwards to:

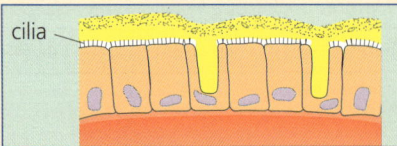

   - A   kill bacteria
   - B   cool down the air in the lungs
   - C   move the mucus to the throat area
   - D   circulate the air in the lungs

5. Cells lining the lungs of smokers become
   covered in tar. Tar is harmful. Which of the
   following problems is **not** true of tar?
   - A   it slows down diffusion of oxygen into
         the blood
   - B   it slows down diffusion of carbon
         dioxide out of the blood
   - C   it causes cancer
   - D   it is addictive

6. After each meal, which hormone results in a
   decrease in glucose in the blood?
   - A   insulin
   - B   glucagon
   - C   adrenaline
   - D   testosterone

7. The diagram below shows a section through a
   red blood cell. Red blood cells can transport a
   lot of oxygen because they:
   - A   have a low surface area
         to volume ratio
   - B   contain haemoglobin
   - C   have a large nucleus
   - D   have a cell membrane

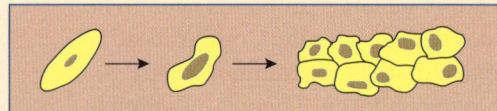

8. The diagram below shows how a lung becomes
   abnormal and increases in number. Which
   disease is shown by the diagram?

   - A   emphysema
   - B   bronchitis
   - C   cancer
   - D   pneumoconiosis

9. The diagram shows a red blood cell from a
   person suffering from a genetic disease. What is
   the disease?
   - A   cystic fibrosis
   - B   sickle cell anaemia
   - C   influenza
   - D   bronchitis

10. In a long distance race we respire aerobically.
    This produces a lot of:
    - A   oxygen
    - B   carbon dioxide
    - C   glucose
    - D   lactic acid

11 The graph below shows the level of glucose in the blood of a diabetic person during one day. The evidence from the graph that identifies the person as diabetic is:

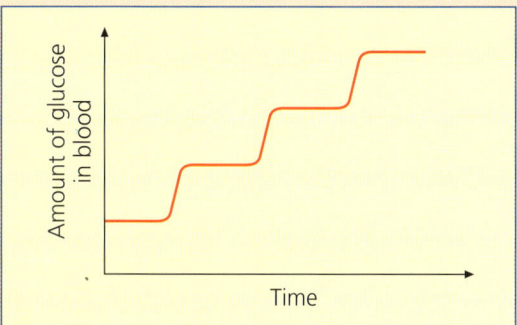

Time

A blood glucose decreases a lot after a meal
B blood glucose remains high after a meal
C blood glucose rises and falls
D blood glucose remains the same

12 The diagram below shows a section through a vein. The valve would open if there was:

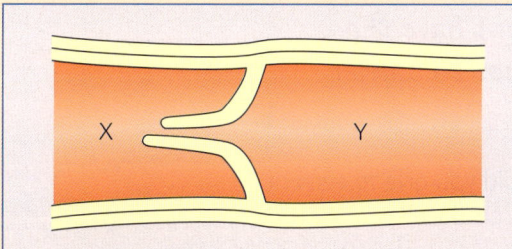

A higher pressure at X and lower pressure at Y
B higher pressure at Y and lower pressure at X
C equal pressure at X and equal pressure at Y
D higher pressure outside the vein and lower pressure inside the vein

13 Gaseous exchange takes place at the surfaces of alveoli. Which line correctly describes which gas or gases enter the blood and which leave?

	From alveoli into blood	From blood into alveoli
A	carbon dioxide + oxygen	water
B	oxygen + water	carbon dioxide
C	oxygen	carbon dioxide + water
D	carbon dioxide + water	oxygen

14 The graph below shows the volume of oxygen used before, during and after exercise. Which letter on the graph (**A**, **B**, **C** or **D**) shows when maximum oxygen debt is being repaid?

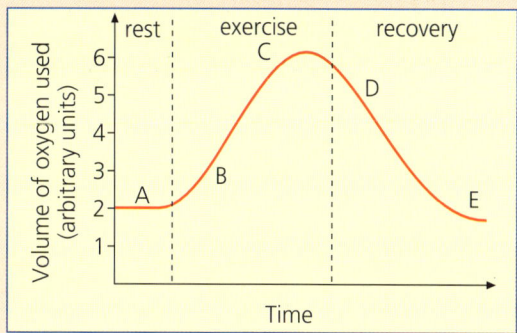

15 The total amount of blood that passed out of the heart in 1 minute was 4800 cm^3. What is the heart rate in beats per minute?

A 800
B 80
C 8
D 0.8

16 Which of the following statements about blood vessels is correct?

A arteries are more muscular than veins, so contract more
B veins are more muscular than arteries, so contract more
C arteries have a lower blood pressure than veins
D capillaries have the highest pressure of all blood vessels

17 Which statement is correct for the action of breathing?

A intercostal muscles lift the rib cage
B intercostal muscles make the rib cage fall
C pressure in the lungs is greater than air outside
D the diaphragm relaxes

18 Emphysema is caused by smoking tobacco. Which of the following statements correctly describes how emphysema develops?

A a build-up of tar reduces the efficiency of gas exchange
B walls of the alveoli begin to break down
C abnormal cells appear and develop into a growth
D chemicals in smoke destroy cilia

Answers to these questions can be found on pages 143–147

Getting it right

1   Cystic fibrosis is a genetic disorder caused by a defective gene. Usually, the lungs and small intestine produce normal mucus, but people with cystic fibrosis secrete thick sticky mucus. The family tree below shows the Parsons family. Use the information in this family tree to answer the following questions.

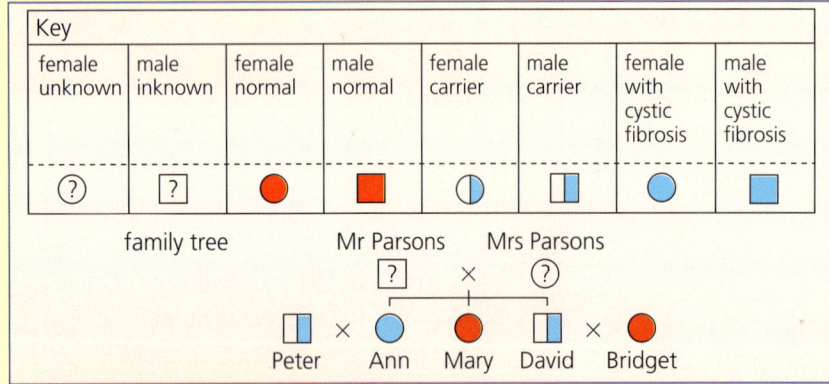

Always look carefully at the information given. Here a key is given. Remember that a carrier is heterozygous.

(a)   Work out the genotypes of Mr Parsons and Mrs Parsons.

> Mary has both dominant alleles so both Mr and Mrs Parsons have the dominant allele. Ann has cystic fibrosis, so both parents must have the cystic allele, so their genotypes must be carrier male and carrier female.

   [3]

(b)   As a genetics counsellor, what information could you give to each of the following couples about children they may produce?
   (i)   Peter and Ann

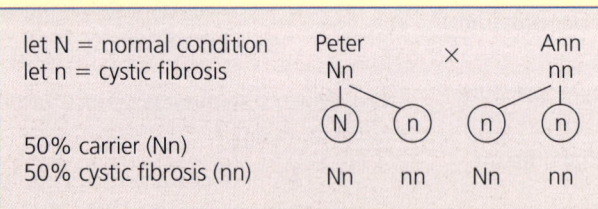

   [2]

Be ready to use your own symbols. Here "N" and "n" were used. Never use letters such as S and s because they are easily confused.

   (ii)   David and Bridget

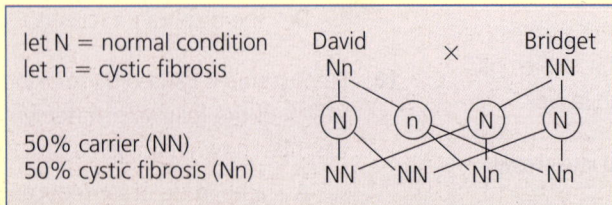

   [2]

In genetics questions you can work forwards or backwards in the family tree. In this question, you need to do both.

Chemicals and the Earth

9

The first topic in this module is about how metals are extracted from their ores. We then look at the metals in the middle of the periodic table – the transition metals. The third topic is on alkali metals, a family of reactive metals in Group 1. We then go on to look at rocks – important building materials and a source of many other materials. In the final topic we look at the atmosphere and the gases in the atmosphere.

Extracting metals from their ores

You need to know ●

✔ about the uses of iron/steel, aluminium and copper;

✔ how metals are extracted from their ores, relating the method to the position of the metal in the order of reactivity;

✔ how iron and aluminium are extracted and how copper is purified by electrolysis.

Uses of metals ● ● ● ● ● ● ● ● ● ● ● ●

The table on the right gives some common uses of iron/steel, aluminium and copper.

Metal	Common uses
iron/steel	ships, cars, bridges
aluminium	aeroplanes, cooking foil, overhead power cables
copper	electrical wiring, water pipes

Extraction of metals ● ● ● ● ● ● ● ●

Most metals are found in the Earth in metal compounds. These are mixed with other minerals in **ores**.

Metals are usually extracted from their ores by **reduction** reactions. Reduction is a reaction in which oxygen is lost. For example, copper(II) oxide is reduced to copper when oxygen is removed:

$$CuO + C \rightarrow Cu + CO$$

Oxidation can also be defined as a process in which electrons are lost. Reduction is a process in which electrons are gained. In the reduction of copper(II) oxide each copper(II) ion gains two electrons to form a copper atom.

Remember
OIL RIG
Oxidation – **L**oss of electrons
Reduction – **G**ain of electrons

The method of extraction required depends upon the position of the metal in the **reactivity series**.

Position of metal in reactivity series	Example	Method of extraction
high	aluminium	electrolysis
middle	iron	reduction with carbon or carbon monoxide
low	gold	metal uncombined in the Earth

Extraction of aluminium

Aluminium is extracted from purified **bauxite** (aluminium oxide) by **electrolysis**. The diagram shows a cell used to extract aluminium.

The electrolyte is aluminium oxide dissolved in molten cryolite (Na_3AlF_6). The lining of the cell is made of carbon and this acts as the cathode (negative electrode). The anodes (positive electrodes) are made of carbon.

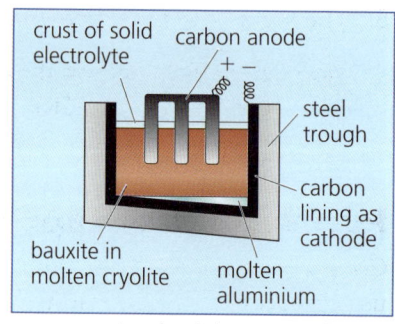

An aluminium extraction cell

At the cathode:

$$Al^{3+} + 3e^- \rightarrow Al$$

At the anode:

$$6O^{2-} \rightarrow 3O_2 + 6e^-$$

Overall the reaction is:

$$2Al_2O_3 \rightarrow 4Al + 3O_2$$

The carbon anodes burn away in the oxygen and have to be replaced.

Extraction of iron ● ● ● ● ● ● ● ● ● ● ● ● ● ● ● ● ●

Iron is extracted from iron ore by reduction in a **blast furnace**. The furnace is loaded with **iron ore**, **coke** and **limestone** and then heated with blasts of hot air.

The products of the furnace are **molten iron** and **slag**. They are tapped off at the bottom of the furnace. Most of the iron is turned into steel. The slag is used as a phosphorus fertiliser and for road chippings.

The main reactions taking place in the furnace are:

1. The burning of coke in air:

carbon + oxygen → carbon dioxide
$$C + O_2 \rightarrow CO_2$$

2. The reduction of carbon dioxide to carbon monoxide:

carbon dioxide + carbon → carbon monoxide
$$CO_2 + C \rightarrow 2CO$$

3. The reduction of iron oxide with carbon monoxide:

iron oxide + carbon monoxide → iron + carbon dioxide
$$Fe_2O_3 + 3CO \rightarrow 2Fe + 3CO_2$$

4. The decomposition of calcium carbonate:

calcium carbonate → calcium oxide + carbon dioxide
$$CaCO_3 \rightarrow CaO + CO_2$$

5. The removal of impurities from the furnace as slag:

calcium oxide + silicon dioxide → calcium silicate
$$CaO + SiO_2 \rightarrow CaSiO_3$$

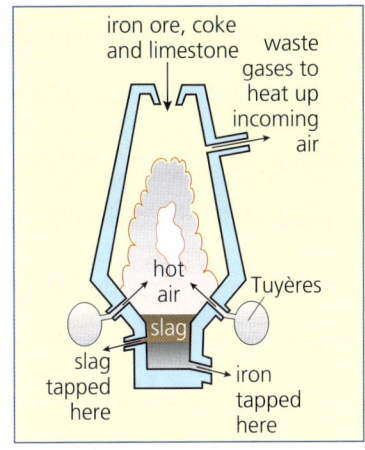

A blast furnace

Reactions 1 and 2 produce the carbon monoxide for the reduction. Reaction 3 is the most important reaction in the furnace producing iron. Reactions 4 and 5 show how limestone removes impurities from the furnace.

Purification of copper by electrolysis

Copper is needed in a very pure form for a number of uses including electrical wiring.

Copper is purified by electrolysis. The impure copper is used to make the anode (positive electrode) in the cell. The cathode (negative electrode) is pure copper. The electrolyte is copper(II) sulphate solution.

During the electrolysis, the anode dissolves and pure copper is deposited on the cathode.

At the anode:

$$Cu(s) \rightarrow Cu^{2+}(aq) + 2e^-$$

At the cathode:

$$Cu^{2+}(aq) + 2e^- \rightarrow Cu(s)$$

The impurities include precious metals such as platinum, which sink to the bottom of the cell and collect as anode mud or slime.

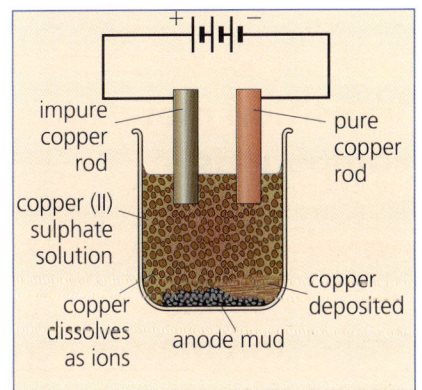

Purification of copper

Questions

1   Which metal in the list could be extracted by electrolysis?
gold lead sodium zinc

2   What three materials are put into the blast furnace in the production of iron?

3   What name is given to a reaction in which oxygen is removed or electrons gained?

Transition metals

Where are transition metals in the periodic table?

Transition metals are found in a block between Groups 2 and 3 in the periodic table.

Transition metals are often able to form more than one positive iron. For example, iron can exist as iron(II) ions and iron(III) ions.

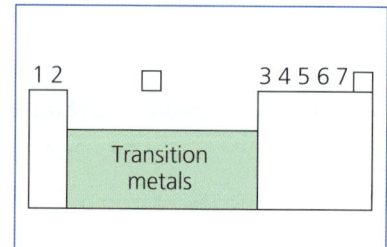

Position of transition metals in the periodic table

Physical properties of iron and copper

Iron and copper have the typical properties of transition metals:

■ high melting points;

■ good conductors of heat and electricity;

■ high density.

Transition metals form coloured compounds

Many transition metal compounds are coloured. For example, copper(II) compounds are blue e.g. copper(II) sulphate. Iron(II) compounds are pale green and iron(III) compounds are red-brown.

Transition metals and their compounds are catalysts

Most common catalysts are either transition metals or transition metal compounds. For example, the catalyst for the decomposition of hydrogen peroxide is manganese(IV) oxide.

You will find other examples in this book of the use of transition metals and their compounds as catalysts.

Questions

The table gives some facts about three metals that have been given the letters A, B and C.

1 Which of the three metals is definitely a transition metal?

2 Which one is definitely not a transition metal?

3 Complete the ionic equation for the conversion of iron(II) into iron(III):

$$Fe^{2+} \rightarrow Fe^{3+}$$

Is this oxidation or reduction?

Metal	Melting point (°C)	Density (g/cm³)	Colour of sulphate
A	64	0.86	colourless
B	420	7.1	colourless
C	1890	7.2	green

Alkali metals

You need to know •

✔ that lithium, sodium and potassium are alkali metals;

✔ that they are found in Group 1 of the periodic table;

✔ that alkali metals have comparatively low melting points and are soft;

✔ about the reactions of alkali metals with water;

✔ how to describe and predict the reactivity of the alkali metals.

Lithium, sodium and potassium are alkali metals. They are found in Group 1 of the periodic table. They have low melting points and are soft. They are stored under paraffin because they are so reactive.

Alkali metal	Symbol	Melting point (°C)	Hard or soft
lithium	Li	180	fairly soft
sodium	Na	98	soft
potassium	K	64	very soft

Outline periodic table showing Group 1

The reactions of alkali metals with water

Alkali metals react with cold water. If a piece of sodium is dropped onto water:

- it floats;
- it melts and a molten ball of sodium is formed;
- it fizzes and shoots about the surface;
- it produces a colourless gas;
- the remaining solution is alkaline (pH>7);

The reaction produces an alkali (sodium hydroxide) and hydrogen gas. The equation for the reaction is:

$$2Na(s) + 2H_2O(l) \rightarrow 2NaOH(aq) + H_2(g)$$

The reactivity of alkali metals increases down the group. With potassium, for example, the hydrogen catches alight and burns with a pinkish-lilac flame

Questions

1 When sodium is added to water, what causes Universal Indicator to turn purple?

2 Complete the word equations below.

Li + H$_2$O → _____ + _____

K + H$_2$O → _____ + _____

3 Rubidium is an alkali metal below potassium in the periodic table. Would you expect it to be more or less reactive than potassium?

Rocks and their uses

You need to know ●

✔ the products of the electrolysis of sodium chloride solution and the uses of these products;

✔ that igneous rocks are produced as the magma cools and crystallises;

✔ that the rate of cooling determines the size of crystals in an igneous rock;

✔ that sedimentary rocks are formed from other rocks and the lower down a layer of sedimentary rock is the older it is;

✔ that fossils can be used to date rocks;

✔ that metamorphic rocks are formed by the action of heat and pressure on existing rocks.

Electrolysis of concentrated sodium chloride solution ● ● ● ● ● ● ● ● ● ● ● ● ● ● ● ●

Electrolysis of concentrated sodium chloride solution produces **sodium hydroxide**, **chlorine** and **hydrogen**.

Sodium chloride solution contains Na^+, Cl^-, H^+ and OH^-. During electrolysis, hydrogen is produced at the cathode and chlorine at the anode:

Cathode: $2H^+ + 2e^- \rightarrow H_2$

Anode: $2Cl^- \rightarrow Cl_2 + 2e^-$

We use sodium hydroxide for making bleach, soap, paper and textile fibres and for drain cleaning. We use chlorine for purification of water supplies, for bleaching and killing bacteria in swimming baths. Hydrogen is used for making margarine and ammonia.

Igneous rocks ● ● ● ● ● ● ● ● ● ● ● ● ● ● ● ● ● ● ●

Igneous rocks, e.g. granite and basalt, are formed when the magma crystallises. Igneous rocks do not contain fossils – this is evidence that these rocks have been formed directly from the magma.

Basalt is formed when the magma crystallises quickly and produces small interlocking crystals.

Granite is formed when the magma crystallises slowly inside the Earth. Granite consists of larger crystals.

Sedimentary rocks ● ● ● ● ● ● ● ● ● ● ● ● ● ● ● ● ●

When weathering and erosion break down existing rocks, sediments are produced. These are transported by rivers into the sea, where they are deposited. These sediments are then compressed and cemented together to form new sedimentary rocks. Sedimentary rocks include limestone and sandstone.

Sedimentary rocks can contain fossils and the nature of the fossils present can be used to date rocks.

In a succession of rocks, for example a cliff face, the older the rock is the lower down it is in the Earth.

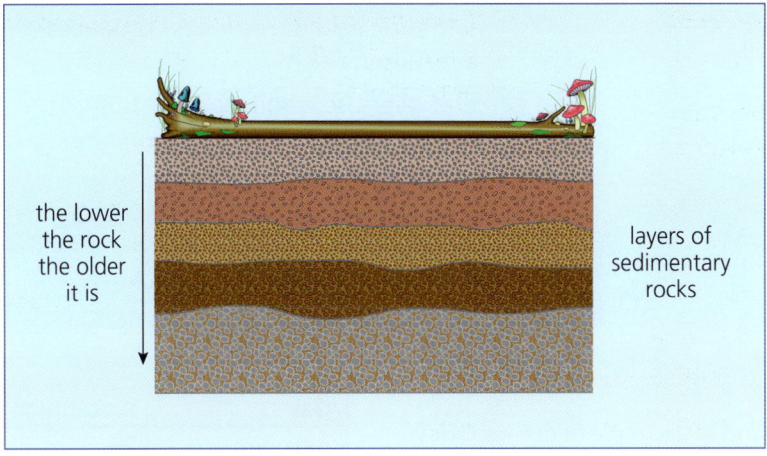

the lower the rock the older it is

layers of sedimentary rocks

Succession of layers of rock

Metamorphic rocks • • • • • • • • • • • • • • • •

The effect of high temperature and high pressure on existing rocks can cause them to change and form new rocks. Marble, for example, is a metamorphic rock produced by the action of high temperatures and high pressures on limestone. The fact that marble and limestone are both forms of calcium carbonate is evidence that marble is made from limestone.

Questions

Of the three types of rock – granite, limestone and marble:

1   Which rock is a sedimentary rock?

2   Which rock is a metamorphic rock?

3   Which rock is an igneous rock?

4   At which place (X, Y or Z) is a metamorphic rock most likely to form?

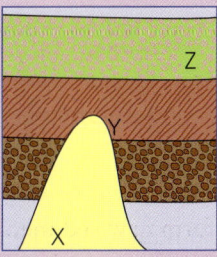

The atmosphere and the gases in it

You need to know •

✔ how and why the composition of the atmosphere has changed;

✔ how the composition of the atmosphere is kept in balance;

✔ the conditions under which nitrogen and hydrogen combine to form ammonia;

✔ that nitrogen fertilisers can promote growth in plants;

✔ that neutralising ammonia with sulphuric acid or nitric acid produces a fertiliser;

✔ that over-use of nitrogen fertilisers can cause environmental problems;

✔ that noble gases are inert because their atoms have full electron shells;

The atmosphere •

The table shows the typical composition of a sample of dry air today.

The atmosphere has changed over millions of years. The table shows how these changes may have taken place.

The composition of the atmosphere is kept constant because some processes use up oxygen (respiration, combustion, rusting) and another (photosynthesis) produces it again.

Gas	Percentage
nitrogen	79
oxygen	20
argon	0.9
carbon dioxide	0.04
plus small amounts of helium and neon	

Change	Effect on the atmosphere
the first atmosphere	consisted mainly of **hydrogen** and **helium**
volcanoes started to erupt	mostly **carbon dioxide** and **water vapour** entering the atmosphere; smaller quantities of **methane** and **ammonia**
Earth cools	water vapour condenses to **liquid water**. Oceans start to form
nitrifying and denitrifying bacteria start to work	ammonia converted to **nitrates** and nitrates are converted into **nitrogen** gas
methane in the atmosphere burns	**carbon dioxide** is formed
photosynthesis occurs	plants convert carbon dioxide into **oxygen**
burning of fossil fuels	increasing levels of carbon dioxide

Using nitrogen in the air to make ammonia •

Nitrogen and hydrogen react together to produce ammonia in the Haber process.

The mixture of gases is passed over an iron catalyst at high pressures at 450°C.

An **equilibrium** is set up and about 10% of the ammonia is formed:

$$N_2 + 3H_2 \rightleftharpoons 2NH_3$$

The ammonia is removed by cooling and liquefying. The unreacted gases are recycled.

Conditions for the Haber process

Increasing the pressure of the nitrogen and hydrogen mixture causes more ammonia to be produced. This is called moving the equilibrium to the right. However, increasing the pressure needs expensive equipment.

> In an equilibrium, the forward and reverse reactions are taking place at the same rate.

The reaction of nitrogen and hydrogen is exothermic.

> **nitrogen + hydrogen ⇌ ammonia + energy**

Lowering the temperature will again move the equilibrium to the right and produce more ammonia. However, lowering the temperature will slow down the reaction too much. A compromise has to be made with as low a temperature as possible and using a catalyst to speed up the reaction.

> A catalyst speeds up a reaction but does not alter the position of the equilibrium. It does not produce more ammonia – just the same amount more quickly.

Fertilisers

Nitrogen compounds can act as **fertilisers** to make plants grow better.

Most plants cannot take in nitrogen from the atmosphere but rely on the nitrogen being 'fixed' in compounds that they take up through their roots.

> Fertilisers often have NPK values. This gives the amounts of the three important elements – nitrogen, phosphorus and potassium.

Two important nitrogen fertilisers are ammonium nitrate and ammonium sulphate. These can be made by neutralising ammonia with nitric or sulphuric acid.

Overuse of fertilisers can cause environmental problems.

> **ammonia solution + nitric acid → ammonium nitrate**
> $NH_3(aq)$ + $HNO_3(aq)$ → $NH_4NO_3(aq)$

- Excess nitrogen compounds are washed into rivers.
- Bacteria in the water convert ammonia into nitrates.

> **ammonia solution + sulphuric acid → ammonium sulphate**
> $2NH_3(aq)$ + H_2SO_4 → $(NH_4)_2SO_4(aq)$

- Nitrates make the water plants grow better.
- These plants prevent sunlight entering, reducing photosynthesis.
- When the plants die, oxygen is used up.
- Eventually, the water contains no dissolved oxygen and so all fish and river life die.

Noble gases

The noble gases are in Group 8 of the periodic table. They are very unreactive compared to other elements. These inert properties are a result of the fact that all noble gas atoms have a full outer electron shell. The atoms have no tendency to gain or lose electrons.

The table gives some uses of the three common noble gases.

Noble gas	Uses
helium	air ships, weather balloons
neon	fluorescent light tubes
argon	electric light bulbs

Questions

1. Which gas is present in the largest amounts in air?
2. Which gas is the most reactive gas in the atmosphere?
3. Why does burning of the rainforests threaten to change the composition of the atmosphere?

Practice module test

You will have 17 minutes
to answer these questions

1 Ammonia is made in the Haber process from:
   A nitrogen and hydrogen
   B nitrogen and oxygen
   C hydrogen and oxygen
   D nitrogen, hydrogen and oxygen

2 The percentage of nitrogen in the atmosphere is approximately:
   A 1%
   B 4%
   C 20%
   D 80%

3 Ammonium sulphate is a fertiliser. It is made by reacting ammonia with:
   A sulphur
   B copper sulphate
   C iron sulphide
   D sulphuric acid

4 Lithium reacts with cold water to form:
   A an acid and hydrogen
   B an acid and oxygen
   C an alkali and hydrogen
   D an alkali and oxygen

5 Sedimentary rocks are produced when fragments of other rocks are:
   A compressed and cemented together
   B cooled
   C heated and compressed
   D melted and crystallised

6 Which process produces oxygen?
   A combustion
   B photosynthesis
   C respiration
   D rusting

7 Of the three types of rock – igneous rocks, metamorphic rocks and sedimentary rocks – in which would you find fossils?
   A sedimentary and metamorphic rocks
   B sedimentary and igneous rocks
   C igneous and metamorphic rocks
   D sedimentary, igneous and metamorphic rocks

8 Sodium hydroxide is used to make:
   A baking powder
   B soap
   C table salt
   D toothpaste

9 Electrolysis of concentrated sodium chloride solution produces:
   A sodium and chlorine
   B sodium hydroxide, hydrogen and chlorine
   C sodium hydroxide, oxygen and hydrogen
   D sodium hydroxide, chlorine and oxygen

10 A reaction that happens in a blast furnace is:
    iron oxide + carbon monoxide
    → iron + carbon dioxide
    Which substance is reduced?
    A carbon dioxide
    B carbon monoxide
    C iron
    D iron oxide

Questions 11 and 12 are about the purification of copper by electrolysis.

11 The substance produced at the cathode (negative electrode) is:
    A copper
    B hydrogen
    C oxygen
    D sulphuric acid

12 The ionic equation for the reaction at the cathode (negative electrode) is:
    A $Cu^{2+} + 2e^- \rightarrow Cu$
    B $Cu^{2+} \rightarrow Cu + 2e^-$
    C $Cu^{2+} + e^- \rightarrow Cu^+$
    D $Cu^+ + e^- \rightarrow Cu$

Questions 13 and 14 are about the Haber process for producing ammonia. The equation for this reaction is

$$N_2 + 3H_2 \rightleftharpoons 2NH_3$$

13 Which of these increases the yield of ammonia in this process?

    A decreasing temperature and increasing pressure

    B increasing temperature and adding a catalyst

    C increasing pressure and adding a catalyst

    D adding a catalyst only

14 What volume of hydrogen is needed to react with 300 dm³ of nitrogen?

    A 100 dm³

    B 300 dm³

    C 600 dm³

    D 900 dm³

Questions 15–17 refer to the section through the Earth's crust.

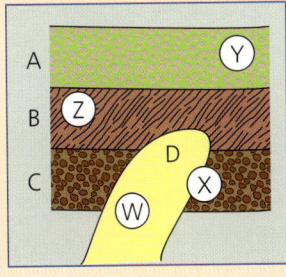

15 Which rock (A, B, C or D) is an igneous rock?

16 Which rock (A, B, C or D) is the oldest sedimentary rock?

17 In which place (W, X, Y or Z) is a metamorphic rock likely to be found?

    A W

    B X

    C Y

    D Z

18 Some metal oxides catalyse the decomposition of hydrogen peroxide. Which metal oxide would most likely catalyse this reaction?

    A Na_2O

    B MgO

    C CaO

    D Fe_2O_3

19 The equation for the reaction of potassium and cold water is:

    A $2K + H_2O \rightarrow K_2O + H_2$

    B $2K + 2H_2O \rightarrow 2KOH + H_2$

    C $4K + 2H_2O \rightarrow 4KH + O_2$

    D $4K + 2H_2O \rightarrow 4KOH + O_2$

20 Helium and argon are unreactive gases because the atoms:

    A contain two electrons in the outer shell

    B contain eight electrons in the outer shell

    C have a full outer shell of electrons

    D have an even number of electrons

Answers to these questions can be found on pages 143–147

Getting it right

1 **(a)** The composition of the atmosphere is kept approximately in balance by several different processes. Explain how these processes do this.

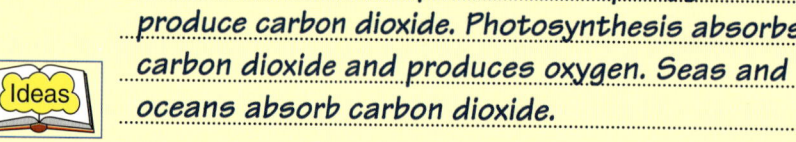

Combustion and respiration use up oxygen and produce carbon dioxide. Photosynthesis absorbs carbon dioxide and produces oxygen. Seas and oceans absorb carbon dioxide. **[4]**

(b) Suggest two environmental changes that might alter the balance.

1. Cutting down forests will reduce photosynthesis. 2. Pollution of the seas might reduce the absorption of carbon dioxide. **[2]**

2 Sodium reacts with cold water. The symbol equation for the reaction is

$$2Na + 2H_2O \rightarrow 2NaOH + H_2$$

(a) (i) Write down an ionic equation for the change from sodium atoms to sodium ions.

$Li \rightarrow Li^+ + e^-$ **[2]**

(ii) Explain why lithium is oxidised in this reaction.

This equation shows that an electron is lost. Loss of electrons is oxidation. **[1]**

(b) (i) Write down the arrangement of electrons in lithium, sodium and potassium atoms.

Lithium 2,1

Sodium 2,8,1

Potassium 2,8,8,1 **[3]**

(ii) Suggest why the reactivity of alkali metals increases down the group.

Down the group the atoms get larger. Outer electrons are further from the nucleus. Force of attraction between nucleus and outer electron is reduced. Electrons are more easily lost down the group. **[4]**

Three marks in (a) are testing Ideas and Evidence. They are dealing with environmental issues. One mark is allocated for QWC. It is awarded if candidates use correct spelling, punctuation and grammar to make the meaning clear.

There may be other acceptable answers. The word 'suggest' means there may be a variety of possible answers.

Understanding oxidation and reduction in terms of loss and gain of electrons is important for Higher level.

(b)(i) is testing Module 3, but it will help you with (b)(ii) The questions in the end of course exams do not just test one module.

Understanding chemical reactions

The first topic in this module is intended to extend your understanding of the structure of the atom and the particles that make up atoms. The second topic looks at the types of forces holding atoms together. The third topic is about energy transfers that take place when reactions occur. In the final topic, reacting quantities are calculated and used to work out chemical formulae.

Atoms and isotopes

Protons, neutrons and electrons • • • • • • •

All atoms are made up of three types of particle – protons, neutrons and electrons. The table below gives information about these three particles.

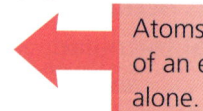

Atoms are the smallest part of an element that can exist alone.

Particle	Mass	Charge
proton, p	1 unit	$+1$
neutron, n	1 unit	0
electron, e	negligible	-1

All the atoms in an element contain the same number of protons. All atoms are neutral.

Atoms must contain equal numbers of protons and electrons.

The protons and neutrons in an atom are tightly packed together in the **nucleus**. The nucleus of an atom is positively charged.

Electrons are arranged in shells around the nucleus.

Each atom has a **mass number** and an **atomic number**. The mass number is the number of protons and neutrons in the nucleus of the atom. The atomic number is the number of protons in the nucleus of the atom.

For example, a sodium atom has a mass number of 23 and an atomic number of 11. This means that the nucleus of a sodium atom contains 11 protons and (23–11), i.e. 12 neutrons. There are 11 electrons around the nucleus.

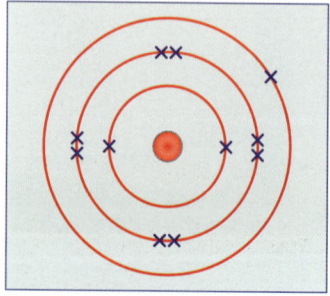

A sodium atom

Isotopes

Hydrogen gas is made up of hydrogen atoms only. But there are three different types of hydrogen atom – they are all hydrogen atoms because they contain one proton and one electron, but they contain different numbers of neutrons.

Atoms of the same element containing different numbers of neutrons are called **isotopes**.

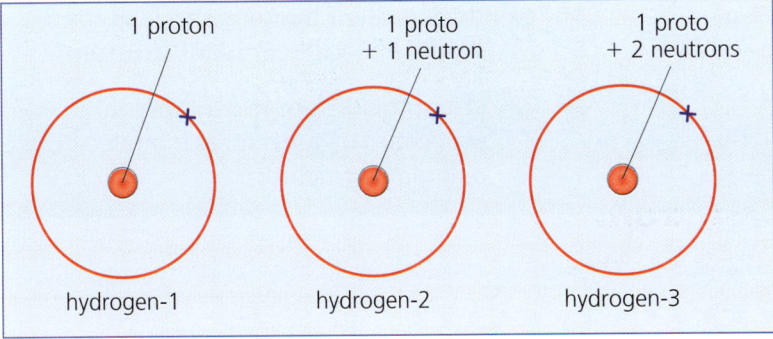

| 1 proton | 1 proto + 1 neutron | 1 proto + 2 neutrons |
| hydrogen-1 | hydrogen-2 | hydrogen-3 |

The three isotopes of hydrogen

> *Students often confuse isotope with isomer and allotrope – make sure you know the difference.*

The average mass number, taking into account the different amounts of the different isotopes present is called the **relative atomic mass**.

> *Look up the relative atomic mass of chlorine in a data book. You will get a value of 35.5.*

You can calculate this. For example, chlorine contains two isotopes – chlorine-35 and chlorine-37. A sample of chlorine contains 75% chlorine-35 and 25% chlorine-37.

$$\text{The relative atomic mass of chlorine} = \left(\frac{75}{100} \times 35\right) + \left(\frac{25}{100} \times 37\right)$$
$$= 26.25 + 9.25$$
$$= 35.5$$

Questions

A phosphorus atom has an atomic number of 15 and a mass number of 31.

1 How many protons does a phosphorus atom contain?

2 How many electrons does a phosphorus atom contain?

3 How many neutrons does a phosphorus atom contain?

Oxygen has two isotopes – oxygen-16 and oxygen-18.

4 Each oxygen isotope has atoms containing the same number of protons and electrons. How many?

5 What is the number of neutrons in each isotope?

The element europium has two isotopes, europium-151 and europium-153. A sample of europium contains 50% of each isotope.

6 What is the difference in the number of neutrons in the two isotopes of europium?

7 What is the relative atomic mass of europium?

Chemical bonds

You need to know •••••••••••••••••••••••••••

✔ that chemical bonding involves transfer or sharing of electrons;

✔ that ionic bonds are formed between metal and non-metal atoms, e.g. sodium and chlorine forming sodium chloride;

✔ that compounds containing ionic bonding have lattice structures;

✔ that covalent bonds are formed between atoms of non-metals to produce simple molecules;

✔ that covalent bonding can lead to simple molecules or giant structures;

✔ about the differences in physical properties of substances in molecules and giant structures.

Ionic bonding ••••••••••••••••••••••••

Atoms of different elements can be joined together by two different types of bonding – ionic and covalent bonding.

Ionic bonding joins metal and non-metal atoms together. It involves a complete transfer of one or more electrons.

Sodium chloride is an example. Sodium chloride is formed when sodium burns in chlorine:

$$2Na + Cl_2 \rightarrow 2NaCl$$

A sodium atom has a single electron in the outer shell and a chlorine atom has seven electrons in the outer shell.

A complete transfer of one electron from a sodium atom to a chlorine atom forms two **ions**. The sodium ion has a positive charge and the chloride ion has a negative charge. Strong electrostatic forces hold the two ions together.

An ion is an atom or group of atoms with a positive or negative charge.

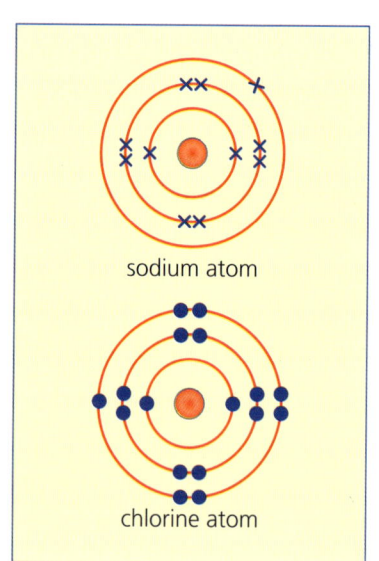

sodium atom

chlorine atom

It is important to stress that a complete transfer of electrons has taken place, from the metal atom to the non-metal atom.

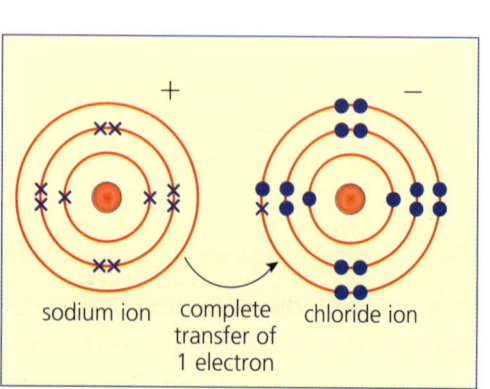

sodium ion   complete transfer of 1 electron   chloride ion

A large number of sodium ions and the same number of chloride ions are joined together in an ionic **lattice**.

The table gives some properties of compounds containing ionic bonding.

Property	Result for ionic compound
state	solid
melting point	high because of strong forces between ions
solubility in water	usually good
solubility in hydrocarbon solvent	poor

Another example of ionic bonding is magnesium oxide.

Covalent bonding

Covalent bonding usually involves joining non-metal atoms together by **sharing** electrons.

The simplest example is a hydrogen molecule, formed when two atoms of hydrogen join together. Each hydrogen atom gives one electron and the pair of electrons is shared to form a **covalent bond**.

Other examples of covalent bonding are chlorine, hydrogen chloride, oxygen and nitrogen.

The table gives some properties of compounds containing covalent bonding.

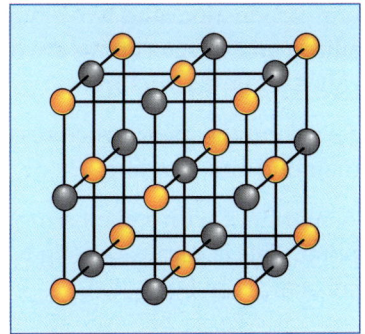

Sodium chloride lattice

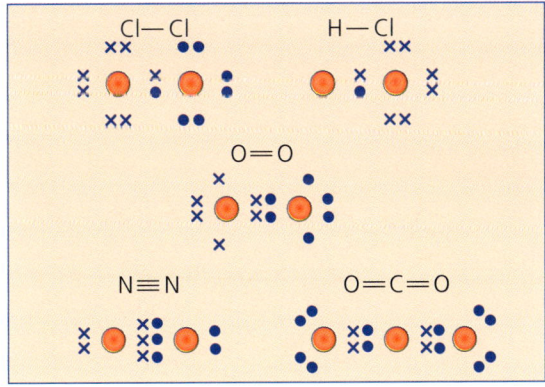

Chlorine, hydrogen chloride, oxygen, nitrogen and carbon dioxide molecules

Property	Result for covalent compound containing molecules
state	usually gas
boiling point	very low
solubility in water	(exception HCl)
solubility in hydrocarbon solvent	good

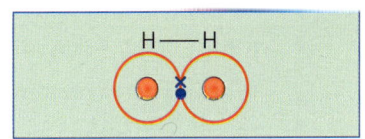

Covalent bonding in hydrogen

continued ⟶

Giant structures

Carbon dioxide and silicon dioxide both contain covalent bonding. However, carbon dioxide is a gas and silicon dioxide is a solid.

Carbon dioxide is made up of separate molecules, each containing one carbon atom and two oxygen atoms. In silicon dioxide all of the silicon and oxygen atoms are linked together in one large molecule, called a **giant structure**.

Two common giant structures are the two forms of carbon – diamond and graphite.

A substance with a giant structure has a high melting and boiling point. This is because strong bonds have to be broken before the substance can be melted.

A substance with a molecular structure has a low melting and boiling point. It is usually a gas, a low melting point solid or low boiling point liquid. Although the bonds within the molecule are strong, there are no forces between molecules.

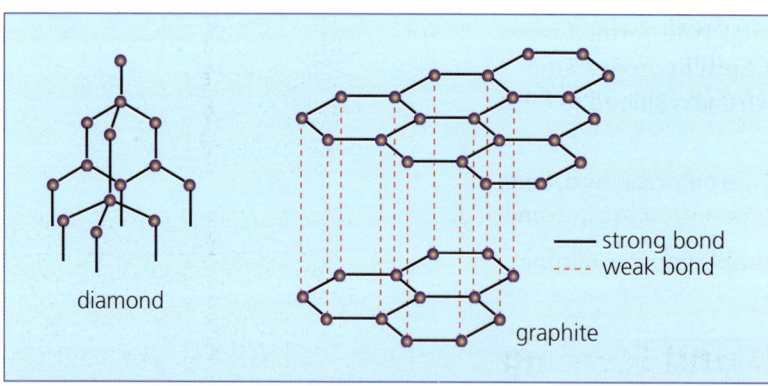

Questions

1   Which type of bonding would be expected in (a) potassium oxide and (b) hydrogen oxide?
2   Which type of bonding involves a complete transfer of electrons?
3   Which type of bonding leads to the formation of molecules?
4   Metal ions are positively charged. True or false?
5   Calcium fluoride, CaF_2, is a high melting point solid. Does it have a giant structure or a molecular structure?

Energy transfers

You need to know •••••••••••••••••••••••••••••••

✔ that temperature changes may be seen during reactions;

✔ that an exothermic reaction is a reaction where energy is given out and an endothermic reaction is a reaction where energy is taken in from the surroundings;

✔ that bond breaking requires energy and bond forming releases energy.

✔ how to work out the relative formula mass of a simple compound using relative atomic masses;

✔ how to use an equation to work out the masses of substances used and produced in a chemical reaction;

✔ how to work out the empirical (simplest) formula for a compound from reacting masses

Exothermic and endothermic reactions

In some reactions there is a rise in temperature during the reaction. For example, when sodium hydroxide and hydrochloric acid are mixed, the temperature rises. This shows that the reaction is giving out energy. It is called an **exothermic reaction**.

When dilute hydrochloric acid is added to a carbonate, there is a slight fall in temperature. The reaction is taking in some energy from the surroundings. This is called an **endothermic reaction**.

> There are few examples of endothermic reactions. At one time scientists thought that endothermic reactions were impossible because they could not find any.

Bonding breaking and bond forming •••

The diagrams show a mixture of hydrogen and chlorine gases reacting to form hydrogen chloride.

When the hydrogen and chlorine react, energy is needed to break the bonds between the two hydrogen atoms in the H_2 molecule and between the two chlorine atoms in the Cl_2 molecule.

When the HCl molecules are formed, energy is released. The reaction is exothermic, i.e. the energy required to break bonds < the energy released when bonds form.

Using chemical equations ••••••••••••

Relative atomic mass

The **relative atomic mass** of an element is the number of times an atom of an element is heavier than the mass of a hydrogen atom. Magnesium has a relative atomic mass of 24. This means a magnesium atom is 24 times heavier than a hydrogen atom.

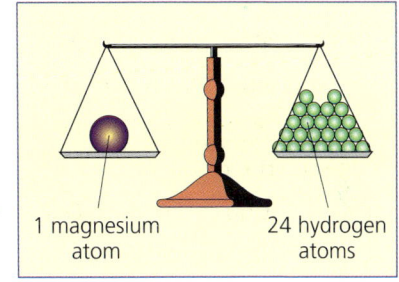

1 magnesium atom   24 hydrogen atoms

> Candidates frequently get this the wrong way round and state that breaking bonds releases energy and forming them requires energy.

continued ➜

Relative formula mass • • • • • • • • • • • • •

The **relative formula mass** of a simple compound is the number of times one molecule of the compound is heavier than the mass of a hydrogen atom.

You can work this out by adding the appropriate relative atomic masses. For example, to find the relative formula mass of water, H_2O:

(Relative atomic masses: O = 16, H = 1)

There are two atoms of hydrogen and one atom of oxygen. So, the relative formula mass of water = $(2 \times 1) + 16 = 18$

You do not have to remember relative atomic masses. They will be given to you when you need them. For example (Relative atomic mass: S = 32, O = 16)

Using chemical equations • • • • • • • • • •

The equation for the reaction of calcium carbonate and hydrochloric acid is:

$CaCO_3(s) + 2HCl(aq) \rightarrow CaCl_2(aq) + CO_2(g) + H_2O(l)$

Using the relative atomic masses we can work out the masses of substances reacting and the masses of substances produced.
(Relative atomic masses: Ca = 40, C = 12, O = 16, H = 1, Cl = 35.5)

Reactants:

$CaCO_3$ RFM = 40 + 12 + (3 × 16) = 100
HCl   RFM = 1 + 35.5 = 36.5

Products:

$CaCl_2$ RFM = 40 +(2 × 35.5) = 111
CO_2 RFM = 12 + (2 × 16) = 44
H_2O RFM = (2 × 1) + 16 = 18

Using the equation:
100 g of calcium carbonate react with 73 g of hydrochloric acid to produce 111 g of calcium chloride, 44 g of carbon dioxide and 18 g of water.

To calculate the mass of hydrochloric acid reacting with 3 g of calcium carbonate:
100 g of calcium carbonate reacts with 73 g of hydrochloric acid

3 g of calcium carbonate reacts with $\frac{73}{100} \times 3$ g of hydrochloric acid

$$= 2.2 \text{ g}$$

To calculate the mass of calcium chloride produced when 3 g of calcium carbonate reacts:
100 g of calcium carbonate produces 111 g of calcium chloride

3 g of calcium carbonate produces $\frac{111}{100} \times 3$ g of calcium chloride

$$= 3.33g$$

A useful check here – the sum of the reactants must always equal the mass of the products. If they do not you have made a mistake – often arithmetical.

Working out the empirical formulae of compounds •

1.12 g of iron combine with 0.48 g of iron oxide.

Divide the masses by the relative atomic masses to get the relative numbers of each atom present in the compound. (Fe = 56, 0 = 16)

Fe	O
$\dfrac{1.12}{56}$	$\dfrac{0.48}{16}$
0.02	0.03

For every two iron atoms there are three oxygen atoms.

The empirical (or simplest) formula of the iron oxide is Fe_2O_3.
The actual formula could be Fe_2O_3, Fe_4O_6, Fe_6O_9, etc.

Questions

1 The reaction for respiration in a cell is

   glucose + oxygen → carbon dioxide + water

   This reaction is exothermic. Finish the equation by adding ' + energy' on the correct side.

2 The equation for photosynthesis is

   carbon dioxide + water → glucose + oxygen

   Is this reaction exothermic or endothermic?

3 Is more energy required to break bonds than to form bonds when methane is burned in air?

The relative atomic mass of carbon and magnesium are 12 and 24.

4 How many times heavier is a carbon atom than a hydrogen atom?

5 How many times heavier is a magnesium atom than a carbon atom?

6 What is the relative formula mass of carbon dioxide, CO_2?
   (Relative atomic masses: C = 12, O = 16)

7 What is the relative formula mass of ammonia, NH_3?
   (Relative atomic masses: N =14, H = 1)

Magnesium reacts with oxygen according to the equation:
   $2Mg(s) + O_2(g) → 2MgO(s)$
   (Relative atomic masses: Mg = 24, O = 16)

8 What mass of oxygen reacts with 4.8 g of magnesium?

9 What mass of magnesium oxide is formed when 4.8 g of magnesium is completely burned in oxygen?

A hydrocarbon contains 6.0 g of carbon and 2.0 g of hydrogen.
   (Relative atomic masses: C = 12, H = 1)

10 What is the empirical formula of this hydrocarbon?

Practice module test

1 Which of the following is an ion?

    A C

    B CO_2

    C CO_3^{2-}

    D CH_4

Questions 2 and 3. Use the table below to answer these questions.

	Atomic number	Mass number
A	14	28
B	13	27
C	15	31
D	12	26

2 Which atom (**A**, **B**, **C** or **D**) contains the most protons?

3 Three of these atoms contain the same number of neutrons. Which one (**A**, **B**, **C** or **D**) contains a different number of neutrons from the others?

4 Which compound does the formula $CuSO_4$ represent?

    A calcium sulphide

    B calcium sulphate

    C copper sulphide

    D copper sulphate

5 The diagram shows the structure of an atom.

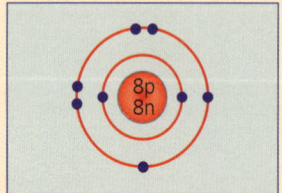

Which row of the table gives the correct atomic number and mass number for the atom?

	Atomic number	Mass number
A	8	8
B	8	16
C	16	8
D	16	16

6 Which row of the table shows a substance with ionic bonding?

	Melting point (°C)	Melting point (°C)
A	−10	60
B	800	2300
C	0	100
D	80	150

7 Ozone is a reactive form of oxygen. It has the formula O_3. An oxygen atom has an atomic number of 8 and a mass number of 16. The relative formula mass of ozone is:

    A 8

    B 16

    C 24

    D 48

Questions 8–10. The table gives information about protons, neutrons and electrons. It is not complete.

Particle	Charge	Relative mass
proton	+1	+1
electron		
neutron		+1

8 What is the relative mass of an electron?

    A negligible

    B +1

    C +2

    D −1

9 What is the relative charge on an electron?

    A −1

    B 0

    C +1

    D +2

10 What is the relative charge on a neutron?

    A −1

    B 0

    C +1

    D +2

11 The molecular formula of sulphur dioxide is SO_2. What is the mass of oxygen combined with 16 g of sulphur?
(Relative atomic masses: S = 32, O = 16)

   A 8 g
   B 16 g
   C 32 g
   D 64 g

12 The elements P and Q react to form a compound. P has seven electrons in the outer shell. Q has two electrons in its outer shell. The formula of the compound is:

   A PQ
   B P_7Q_2
   C QP_2
   D PQ_2

13 What is the relative formula mass of potassium hydrogencarbonate, $KHCO_3$?
(Relative atomic masses: K = 39, H = 1, C = 12, O = 16)

   A 68
   B 76
   C 88
   D 100

14 What is the relative formula mass of lead(II) nitrate, $Pb(NO_3)_2$?
(Relative atomic masses: Pb = 207, N = 14, O = 16)

   A 237
   B 269
   C 317
   D 331

15 Aluminium can be extracted from aluminium oxide. The equation is:

   $$2Al_2O_3 \rightarrow 4Al + 3O_2$$

What mass of aluminium can be obtained from 5.1 tonnes of aluminium oxide?
(Relative atomic masses: Al = 27, O = 16)

   A 2.7 tonnes
   B 5.4 tonnes
   C 10.8 tonnes
   D 108 tonnes

Questions 16–18. Use the table below.

	Melting point	Electrical conductivity	Solubility in water	Appearance
A	high	good when molten	good	white crystals
B	high	good when solid	poor	shiny
C	low	non-conductor	insoluble	white solid
D	high	good when solid	insoluble	dull black

16 Which substance (**A**, **B**, **C** or **D**) has an ionic lattice?

17 Which substance (**A**, **B**, **C** or **D**) has a molecular structure?

18 Which substance (**A**, **B**, **C** or **D**) is a non-metal with a lattice of atoms?

19 Gallium can exist as two isotopes – gallium-69 and gallium-71. A sample of gallium contains 60% gallium-69 and 40% gallium-71. The relative atomic mass of gallium is:

   A 69.0
   B 69.8
   C 70.2
   D 71.0

20 A carbon atom has four outer electrons(x) and an oxygen atom has six outer electrons(•). Which of the following (**A**, **B**, **C** or **D**) shows the outer electrons in carbon dioxide?

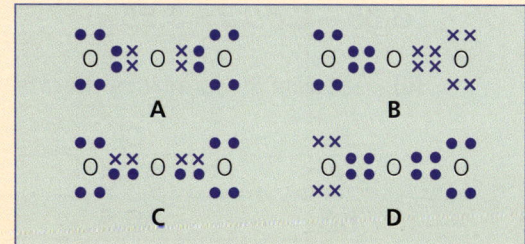

Answers to these questions can be found on pages 143–147

Getting it right

1 Magnesium and oxygen combine to form magnesium oxide.

	Electron arrangement
Magnesium	2,8,2
Oxygen	2.6

(a) Describe the electronic changes that take place when magnesium and oxygen combine.

Magnesium atom loses two electrons; oxygen gains two electrons.   **[2]**

(b) The table gives some information about magnesium oxide and sodium chloride. Why is the melting point of magnesium oxide much higher than that of sodium chloride?

	Melting point (°C)	Ions present
magnesium oxide	2830	Mg^{2+}, O^{2-}
sodium chloride	801	Na^+, Cl^+

Ions in magnesium oxide have 2+ and 2− charges.
Stronger forces of attraction between ions.
Higher temperature needed to break these forces.   **[3]**

2 **(a)** The diagram shows how bonding changes when hydrogen and chlorine react to form hydrogen chloride. Explain how the changes in bonding result in an exothermic reaction.

Energy required to break H−H and Cl−Cl bonds.
Energy released when H-Cl bonds form. More energy released than used.   **[4]**

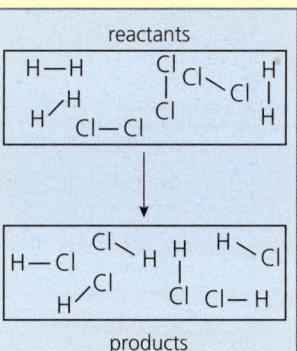

(b) Dry hydrogen chloride reacts with iron to form an iron chloride. 2.80 g of iron produces 6.35 g of iron chloride.
  (i) What mass of chlorine reacts with 2.8 g of iron?

6.35 − 2.80 g = 3.55 g   **[1]**

  (ii) Calculate the formula of the iron chloride.

Fe	Cl
2.80	3.55
56	35.5
0.05	0.1
Formula	$FeCl_2$

[4]

Movement and change

The first topic in this module considers how an object's motion depends on the forces acting on it. It includes a brief study of how our understanding of forces developed through the work of Galileo Galilei and Isaac Newton.

The second topic deals with energy and work and how they are related to each other and to power.

The two final topics are short. One is concerned with seismic waves and how changes in the geology of the Earth happen; the other examines the nature of radioactive decay. Knowledge about how radioactive materials decay is used to date rocks and other very old objects.

Forces and motion

Speed, velocity and acceleration ● ● ● ● ● ● ●

You need to understand that

■ The **speed** (in m/s) of an object is the distance (in m) that it travels each second (s).

■ **Velocity** (in m/s) describes both the speed of an object and its direction of travel.

Speed is shown by the slope of a **distance–time graph**; the steeper the slope, the faster the speed.

Speed can be calculated using the relationship:

speed = distance travelled ÷ time taken or $v = d/t$

Another useful graph is a **speed–time graph**, which shows how the speed of an object changes. When the speed increases, the object is **accelerating**. The steeper the slope of the graph, the greater the acceleration. Acceleration means the increase in speed per second. It is calculated using the relationship:

acceleration = change in speed ÷ time taken

Acceleration is measured in m/s².
A speed–time graph can be used to:

■ calculate acceleration by working out the slope;

■ calculate the distance travelled by finding the area between the curve and the time axis.

A distance–time graph shows how the total distance travelled changes with time. Remember, the distance travelled can only increase, it can never decrease.

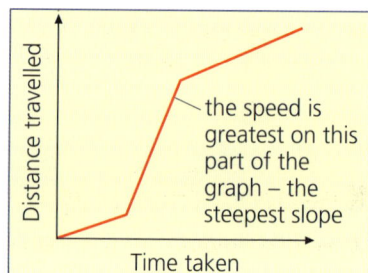

the speed is greatest on this part of the graph – the steepest slope

A distance–time graph

Take care with the unit of acceleration. It means "metres per second each second". In examinations, candidates often lose marks by giving the unit as m/s instead of m/s².

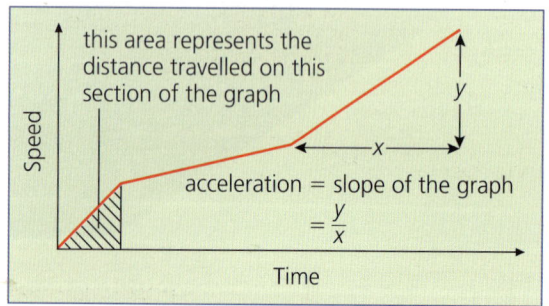

this area represents the distance travelled on this section of the graph

acceleration = slope of the graph
$= \dfrac{y}{x}$

A speed–time graph

Stopping and starting ● ● ● ● ● ● ● ● ● ● ● ● ●

If a driver notices a hazard in the road she must brake. While she is reacting and braking, the vehicle keeps moving.

The total distance that the vehicle travels from when the hazard is seen to when the vehicle stops is called the **stopping distance**. It is made up of:

■ **thinking distance** – the distance that the vehicle travels while the driver reacts;

■ **braking distance** – the distance that the vehicle travels during braking.

Thinking distance depends on:

■ The driver's reaction time – this is longer if the driver has taken drugs or alcohol or is being distracted.

■ The speed of the vehicle – doubling the speed doubles the thinking distance.

Braking distance depends on:

■ The mass of the vehicle – the greater the load, the greater the distance needed to brake.

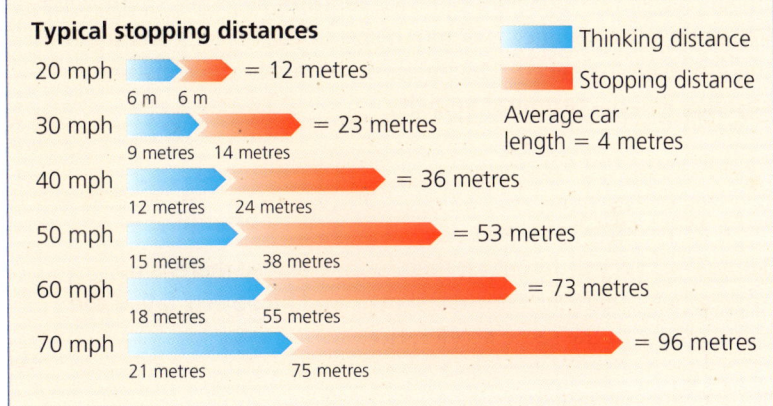

■ The speed of the vehicle – doubling the speed means that it travels four times the distance during braking.

Keeping moving ● ● ● ● ● ● ● ● ● ● ● ● ● ● ●

It seemed obvious to the Greeks that to keep an object moving, you need to keep pushing or pulling it. However, they were not aware of the "unseen" forces of:

■ **friction** – a force that acts against solid objects slipping and sliding over each other;

■ **air resistance** – a force that acts against motion through the air.

friction opposes the motion of the child down the slide

friction force

When you move through a fluid such as air or water, you have to push the fluid out of the way to make room for your body. The force acting against your motion is air resistance or water resistance.

Both Galileo and Newton realised that if an object is not moving or is moving in a straight line at a constant speed then the forces on it are **balanced**. If the forces are balanced, equal forces act in opposite directions to cancel each other out. If the forces on an object are **unbalanced**, it accelerates in the direction of the greater force.

The relationship between the mass of an object and the acceleration that an unbalanced force acting on it causes is:

force = mass × acceleration or $F = m \times a$

continued ⟶

Falling down ● ● ● ● ● ● ● ● ● ● ● ● ● ● ● ● ●

Two forces act on an object that is falling through the air:

■ the Earth's gravitational pull acts downwards – this force is the object's weight;

■ air resistance pushes upwards – the size of this force depends on the shape of the object and its speed.

As an object falls, the balance of the forces acting on it changes.

At low speeds there is little air resistance and the weight of the object causes it to accelerate downwards. As it speeds up, the air resistance increases. The acceleration becomes smaller. When the air resistance is equal in size to the object's weight, the forces are balanced. The object now travels at a constant speed called the **terminal velocity**.

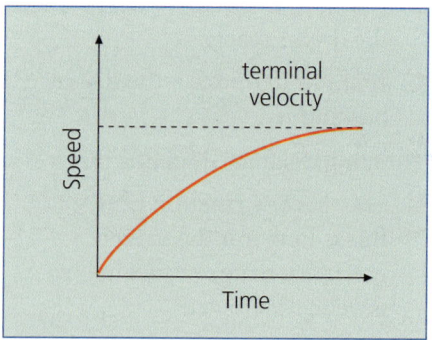

On the Moon ● ● ● ● ● ● ● ● ● ● ● ● ● ● ● ●

Things are very different on the Moon. Because there is no atmosphere, there is no air resistance, and all objects fall at the same rate with the same acceleration.

Back to Earth ● ● ● ● ● ● ● ● ● ● ● ● ● ● ● ● ●

Things fall downwards because of the gravitational attraction between them and the Earth. Newton realised that gravitational attractive forces act between all objects but they are often too small for their effects to be noticed. The forces are only large enough to make things move when one of the objects has a lot of mass, as planets have.

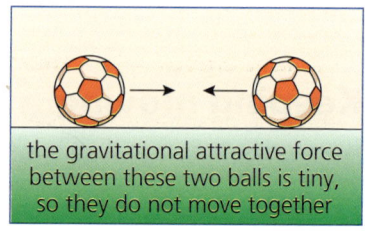

the gravitational attractive force between these two balls is tiny, so they do not move together

Newton also realised that objects exert forces on each other. This means that if you kick a ball the force on the foot and the force on the ball are of **equal size and act in opposite directions**. The forces also have different effects because they act on objects of different mass. The ball has a small mass and so it gains a large forward-directed acceleration. The large mass of the person means that they gain only a small acceleration in a backwards direction.

Questions

1 An object is falling at a constant speed through the Earth's atmosphere. What is this speed called?

2 A car speeds up from 15 m/s to 30 m/s in 12 s. Calculate its acceleration.

3 Explain why "doubling the speed doubles the thinking distance".

Forces and energy

Working

When you lift something up or push something along, you are working. **Work** is being done when a force makes something move.

Work is measured in joules (J) and the amount of work done can be calculated using the relationship:

work done = force × distance moved or $W = F \times d$

> You are working even when you are sat perfectly still, because the cells in your body are respiring and your heart is pumping blood around your body.

Work and energy

To be able to do work, you need **energy**. As you work, energy flows from you to the object being moved.

Two important ways in which objects can gain energy are:

- through movement – a moving object can do work and is said to have **kinetic energy**, which can be calculated using the relationship:

kinetic energy = $\frac{1}{2}$ × mass × (speed)^2 or $E_k = \frac{1}{2}mv^2$

- through position – an object that has been lifted up stores energy as **gravitational potential energy** (gpe), which can be calculated using:

gpe = mass × gravitational field strength × vertical height or gpe = $m \times g \times h$

> People get energy from the food that they eat.

> A moving ball or vehicle has kinetic energy. Energy stored in the weight of a grandfather clock is gravitational potential energy.

> *You need to be able to recall these relationships*

How powerful?

Power describes the amount of work done or energy transferred each second. It is calculated using the relationship:

power = work done ÷ time taken or $P = W \div t$

Power is measured in watts (W), where one watt is equivalent to one joule per second; 1 W = 1 J/s.

Questions

1 A man uses a force of 30 N to push a supermarket trolley 150 m. How much work does he do?

2 A ball is thrown up into the air. What energy transfer takes place immediately after it leaves the hand?

3 Calculate the kinetic energy of a 850 kg car travelling at 14 m/s.

Earth waves

Waves that travel through the body or along the surface of the Earth are called **seismic waves**. They can be caused by earthquakes or by an explosion occurring underground.

The detection of seismic waves provides information about changes taking place in the structure of the Earth.

The outer layer of the solid part of the Earth is called the **lithosphere**. The diagram shows how it is made up of several sections called **plates**. Convection currents inside the Earth cause the plates to move.

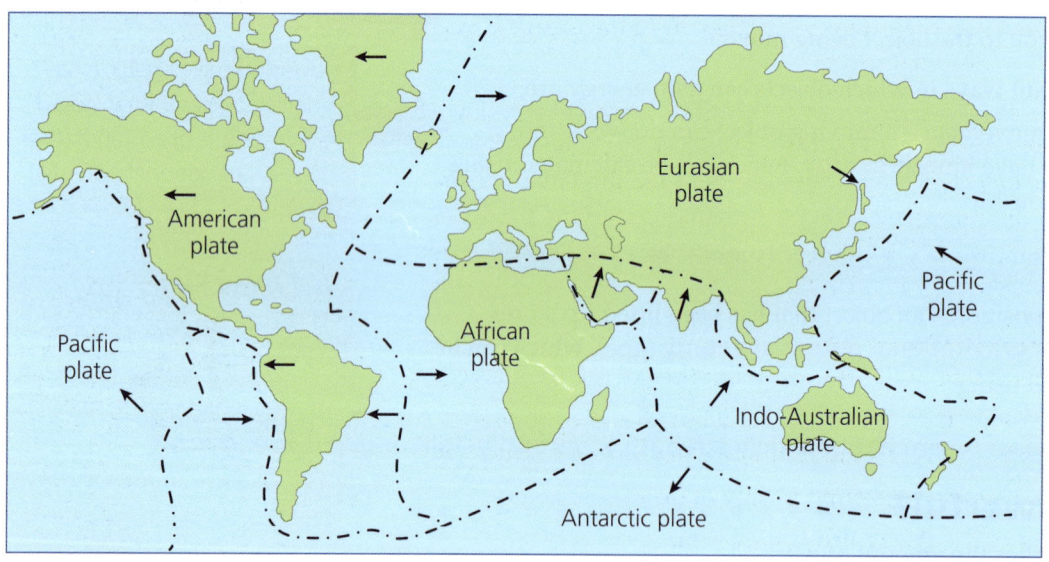

The plates that make up the surface of the earth

The moving plates can:

- slide past each other – an earthquake is caused when there is a sudden "jolt";
- move towards each other – this can result in the formation of mountain ranges;
- move away from each other – this causes volcanoes.

When plates move towards each other, the ocean floor plate is forced under the continental crust plate. Rock from the ocean floor plate is heated and melts in the Earth's **magma**, the hot molten rock in the mantle underneath the Earth's crust.

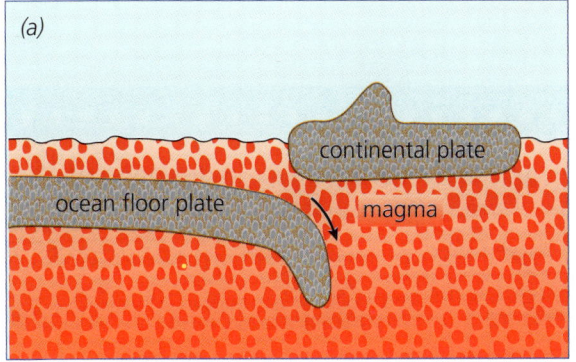

(a)

continental plate

ocean floor plate   magma

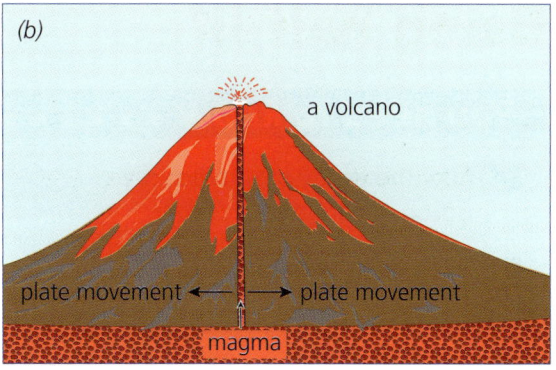

(b)

a volcano

plate movement ← → plate movement

magma

Plates moving (a) towards each other, (b) away from each other

When plates move away from each other, molten rock from the magma escapes from the gap between them. This magma cools and solidifies to form new rock. If this occurs on dry land it forms a volcano.

Two types of seismic waves that travel through the Earth following an earthquake are:

P-waves are the first to be detected as they have the greatest speed. P-waves are longitudinal waves.

S-waves are detected after P-waves because they travel at a slower speed. S-waves are transverse. The diagram shows where these waves can be detected. No S-waves are detected directly opposite the centre of the earthquake. This shows that part of the Earth's core must be liquid, because longitudinal waves can travel through liquid but transverse waves cannot.

The solid Earth has four layers:

- crust – this is thin under the ocean but thicker under mountains;
- mantle – hot rock that is mainly solid but allows very slow-moving convection currents that move the plates;
- liquid outer core – consisting mainly of iron;
- solid inner core – this is also mainly iron.

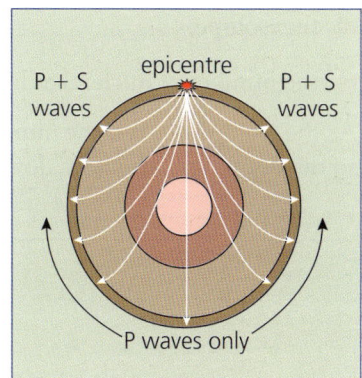

P + S waves   epicentre   P + S waves

P waves only

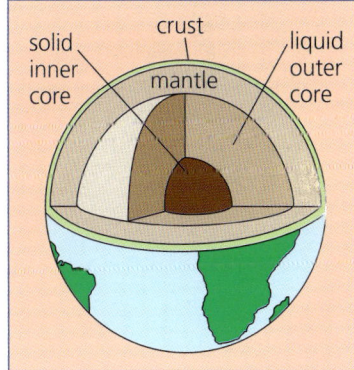

solid inner core   crust   liquid outer core

mantle

Questions

1 What type of plate movement causes a volcano?

2 After an earthquake, what type of wave is detected first?

3 Explain why S-waves are not detected at the surface of the Earth directly opposite the centre of an earthquake.

Using half-life

You need to know •

✔ how the decay of a radioactive isotope changes over time;

✔ the meaning of half-life;

✔ about the problems in disposing of radioactive waste;

✔ how radioactivity is used to date some objects.

Some isotopes are stable – others are unstable. The radioactive decay of an isotope is measured in **becquerels** (Bq), where 1 Bq = 1 decay per second. This decay rate depends on:

■ the isotope;

■ the number of nuclei that have not yet decayed.

As the isotope decays, the number of undecayed nuclei goes down, so the decay rate also goes down.

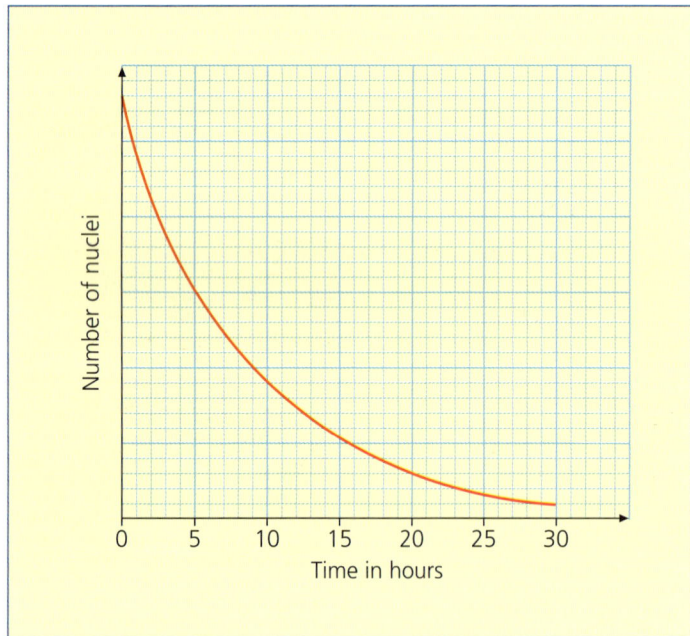

No matter what the time-scale, all radioactive decay follows the pattern shown in the graph.

For the isotope shown in the graph, the number of remaining, undecayed nuclei halves every 6 hours. This time is called the **half-life** of the isotope.

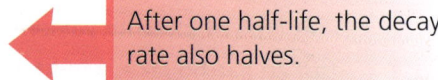

After one half-life, the decay rate also halves.

Some isotopes have much longer half-lives and some have much shorter ones, but they all show the same pattern of decay.

Radioactive dating

Whatever the number of nuclei of a radioactive isotope present in a sample of radioactive material, half of them decay during each half-life. This means that:

- after one half-life has passed, half of the nuclei are undecayed;
- after n half-lives have passed, the number of undecayed nuclei is $\frac{1}{2}^n$.

> A common wrong idea at GCSE is that after two half-lives "all the nuclei have died". Nuclei do not die, they decay. After two half-lives, one-quarter of them are undecayed.

Worked example

Q   Edexcelium has a half-life of 200 years. If a sample contains 1000 radioactive nuclei in the year 2000, how many undecayed nuclei will remain in the year 3000?

A   Number of years elapsed = 3000 − 2000 = 1000 years

Number of half-lives = $\dfrac{1000}{200}$ = 5 half-lives

Number of undecayed nuclei = $\frac{1}{2}^5$ = $\times \frac{1}{2} \times \frac{1}{2} \times \frac{1}{2} \times \frac{1}{2} \times \frac{1}{2}$

= $\frac{1}{32} \times 1000$ = 31.25 = approx 31 undecayed nuclei will remain.

Measuring the number of nuclei of an isotope present in an object can be used to estimate its age:

- Living things absorb small amounts of radioactive carbon-14 from the air and their food. When a living thing dies, the amount of radioactive carbon-14 that it contains decays with a half-life of 6000 years. Measuring the amount of carbon-14 remaining enables the date of death to be estimated.

> Radiocarbon dating is a useful technique for dating old timber and archaeological samples.

- Rocks contain radioactive isotopes. The fewer the nuclei of these isotopes present in a rock, the greater its age.

> Radioactive dating of rocks has been used to estimate the age of the Earth and the Moon.

Knowing the half-life of radioactive materials also helps in deciding how to dispose of radioactive waste. Waste must be stored until ten half-lives have passed for it to be "safe".

For some isotopes used in medicine this takes only a few days, but for some waste materials from nuclear power stations this can take thousands of years.

Questions

1. A radioactive isotope has a half-life of 6 hours. Its rate of decay is measured to be 1200 Bq. Calculate the rate of decay (a) after 6 hours; (b) after a further 6 hours.

2. Which of the following can be dated using radiocarbon dating?
   cloth   pottery   wood

3. The activity of a radioactive sample is measured to be 240 Bq; 90 seconds later it has fallen to 30 Bq. Calculate the half-life of the isotope.

Practice module test

1 Velocity describes:

   A speed and its unit

   B speed upwards

   C speed downwards

   D speed and direction

Questions 2–5

The diagram show the horizontal forces acting on a car.

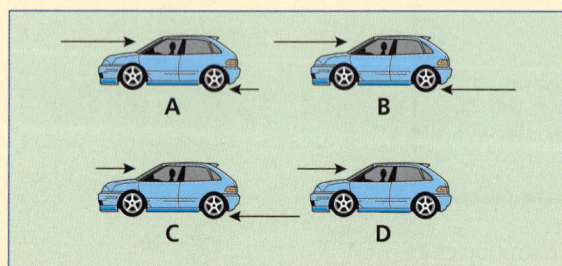

2 Which car is accelerating forwards? *C*

3 Which car has balanced forces acting on it? *B*

4 Which car could be braking? *A*

5 Which car is travelling at a constant speed? *B*

6 A car speeds up from 12 m/s to 16 m/s in 2.0 s. Its acceleration is:

   A 0.5 m/s

   B 0.5 m/s^2

   C 2.0 m/s

   D 2.0 m/s^2

7 Power means:

   A the size of a force

   B the strength of a force

   C how much work is done

   D the work done by a force per second

8 When a weightlifter lifts some weights, the weights gain:

   A gravitational potential energy

   B kinetic energy

   C movement energy

   D thermal energy

9 The outermost layer of the solid Earth is called:

   A the duosphere

   B the ionosphere

   C the lithosphere

   D the monosphere

10 When an ocean plate moves under a continental plate, rocks are:

   A melted in the Earth's inner core

   B melted in the Earth's outer core

   C melted in the Earth's crust

   D melted in the Earth's mantle

11 The half-life of a radioactive isotope:

   A increases with increasing time

   B decreases with increasing time

   C has a constant value

   D is half of the dead-life

12 A radioactive isotope has a half-life of 60 s. The decay rate is measured to be 120 Bq.
120 s later the decay rate is likely to be:

   A 30 Bq

   B 60 Bq

   C 240 Bq

   D 480 Bq

Questions 13 to 15

The graph represents the motion of a cyclist on a short journey.

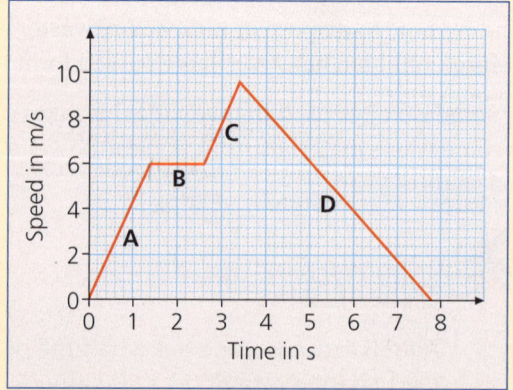

13 The cyclist was stationary for:

   A 0.8 s

   B 0.9 s

   C 1.0 s

   D 1.1 s

14 The cyclist's acceleration on section A of the graph is:

    A 0.25 m/s

    B 0.25 m/s^2

    C 4.0 m/s

    D 4.0 m/s^2

15 In section D of the graph the cyclist travelled a distance of:

    A 0.45 m

    B 2.23 m

    C 21.12 m

    D 43.1 m

16 A car of mass 900 kg accelerates at 1.2 m/s^2. The size of the unbalanced force on the car is:

    A 750 J

    B 1080 J

    C 750 N

    D 1080 N

17 When a driver goes on holiday the mass of the car increases. This causes:

    A the thinking distance to increase

    B the thinking distance to decrease

    C the braking distance to increase

    D the braking distance to decrease

18 A 3.0 N force is used to accelerate a mass of 2.0 kg. The acceleration is:

    A 1.5 m/s

    B 6.0 m/s

    C 1.5 m/s^2

    D 6.0 m/s^2

Questions 19–21

A 0.40 kg ball is thrown vertically upwards with an initial speed of 5.0 m/s. The value of the Earth's gravitational field strength is 10 N/kg.

19 Its initial kinetic energy is:

    A 0 J

    B 0.4 J

    C 1.0 J

    D 5.0 J

20 At its maximum height its kinetic energy is:

    A 0 J

    B 0.4 J

    C 1.0 J

    D 5.0 J

21 The maximum height that the ball reaches is:

    A 1.25 m

    B 2.50 m

    C 5.0 m

    D 12.5 m

22 Which statement about seismic waves is correct?

    A S-waves and P-waves both travel through liquids and solids

    B neither S-waves nor P-waves can travel through solids

    C S-waves cannot travel through solids

    D S-waves cannot travel through liquids

23 Uranium-238 decays into lead-206 with a half-life of 4500 million years. A rock sample contains uranium-238 and lead-206 in equal proportions. The age of the rock is likely to be:

    A 2250 million years

    B 4500 million years

    C 9000 million years

    D 18 000 million years

24 Radioactive carbon, carbon-14, decays with a half-life of 5730 years. It cannot be used to date an archaeological specimen thought to be:

    A 200 years old

    B 2000 years old

    C 5730 years old

    D 10 000 years old

Answers to these questions can be found on pages 143–147

Getting it right

An astronaut drops a piece of Moon rock. The graph shows how the speed of the rock changes from when it leaves the astronaut's hand to when it reaches the Moon's surface.

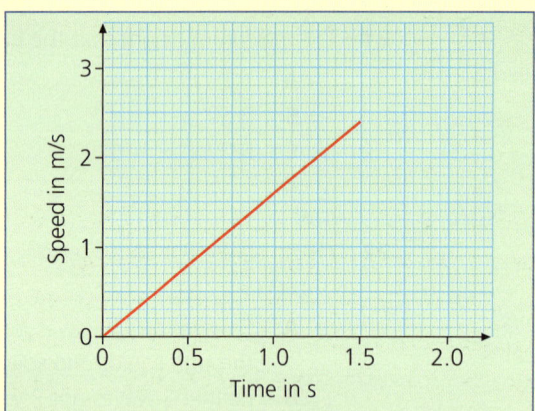

(a) Calculate the acceleration of the rock.

Acceleration = increase in speed ÷ time

= 2.4 m/s ÷ 1.5 s = 1.6 m/s² **[3]**

(b) How far did the rock travel?

Distance travelled = areas between graph line and time axis

= ½ x 2.4 x m/s x 1.5 s = 1.8 m **[3]**

(c) When a similar rock dropped above the surface of the Earth, its acceleration was greater. Explain why.

The Earth's gravitational field is stronger, so there is a greater unbalanced force on the mass. **[3]**

(d) When Galileo dropped two masses from the top of the leaning tower of Pisa, the heavier mass reached the ground a fraction of a second before the lighter one. Some people claimed that this proved that heavier objects fall faster than lighter ones. Suggest an alternative explanation.

The masses were affected by air resistance. This had a greater effect on the lighter mass than on the heavier one. **[2]**

The units in the second line are not essential, but it's a good idea to always give the units with any physical quantity.

Remember, the cue word "explain" means that you need to give a reason. In this case the reason for the greater force is the increased gravitational field strength.

One key aspect of the assessment of 'idea and evidence' is whether you understand that experimental evidence can be interpreted in different ways.

Energy communication and force

The major topic in this module talks about using electricity to transfer energy. It starts with a study of electrostatics and then links that to current and energy transfer in circuits. This is followed by a study of the generation and transmission of electricity. The communication topic covers wave behaviour and how this is used in effective communications. The final topic is concerned with how forces change the size, shape and volume of materials and how the pressure of a gas depends on its volume.

Charge and energy

Charging and discharging ●●●●●●●●●●

Insulators such as plastics are easily charged. You can easily charge insulators such as plastics, for example by rubbing them with a duster. This results in the transfer of negatively charged **electrons**. The object that gains electrons becomes negatively charged. The object that loses electrons becomes positively charged.

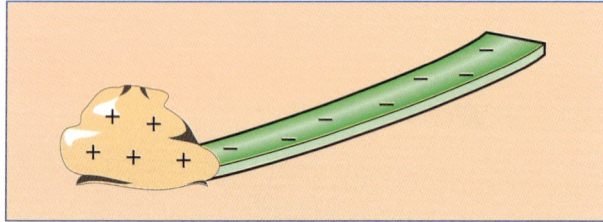

Charging a polythene rod with a duster

Conductors allow electric charge to pass through them – they need to be well insulated from the ground to hold on to static charge.

The balloons in the diagram are held apart by electrostatic forces. The forces are repulsive because both balloons have the same type of charge.

A large build-up of charge can create a high voltage, big enough to make the air conduct electricity. This is what happens when lightning occurs – the high voltage between the bottom of a cloud and the ground splits the air up into positive and negative ions. Movement of these ions results in a flash of lightning and discharges the cloud.

Movement of charge can also give you unpleasant shocks:

- Synthetic fabrics such as nylon and polyester are very good electrical insulators.

- If you sit on a car seat or walk across a nylon carpet while wearing these fabrics, you become charged.

- Touching a car door or an earthed object such as the screw of a light switch causes you to discharge as electrons move between you and the Earth.

- The result is that you feel a shock as charge flows through you.

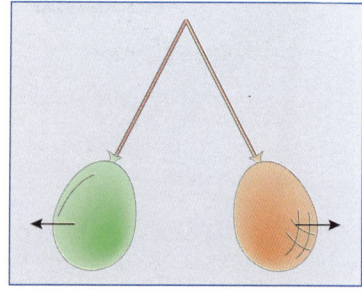

Both balloons have the same type of charge

Charged objects exert electrical forces on each other – like charges repel and unlike charges attract.

When answering questions about electrostatic charge, always use this rule to give the reason for your answer.

Electrostatic charge can be both dangerous and useful, depending on the circumstances:

- when cars and aircraft are being refuelled, charge can build up on the metal bodywork;
- a spark could cause the fuel to ignite.

To avoid this hazard:

- The body of an aircraft is always **earthed** (connected to the ground) while it is being refuelled. This allows movement of electrons to discharge from the aircraft if a voltage develops between the aircraft and the ground.
- The fuel pipe of a car is connected to the metal car body so that the charge spreads out and a high voltage does not build up.

> If the aircraft is connected to the earth, there can be no voltage between them to cause a spark.

Electric charge is useful:

- in photocopiers, where the black powder only sticks to charged areas of the belt;
- in inkjet printers, where the motion of the drops is controlled by charged metal plates.

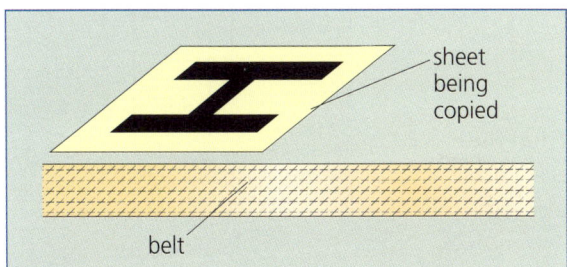

A photocopier uses black powder on a belt to make a print. Where light passes through the sheet being copied, the belt discharges and does not attract the black powder

Questions

1. A balloon is charged with a duster. Electrons move from the duster to the balloon. What type of charge is gained by:

   a the balloon?   b the duster?

2. How does earthing an aircraft prevent an explosion during refuelling?

3. Explain how connecting the fuel pipe of a car to the metal body reduces the risk of a spark occuring.

Charge and current

You need to know •

✔ the relationship between current and charge flow;

✔ the relationship between power, current and voltage;

✔ how to explain the principles of a motor and a transformer;

✔ the relationship between the input and output voltages and the number of turns of wire on transformer coils.

Charge and current • • • • • • • • • • • • • •

Charge needs to be carried through a conductor.

■ In a metal the charge is carried by electrons from positive to negative.

■ In molten or dissolved electrolytes, charge is carried in both directions by ions.

Electric **charge**, Q, is measured in **coulombs** (C).
Any flow of electric charge is a **current**. The relationship between current, charge flow and time is:

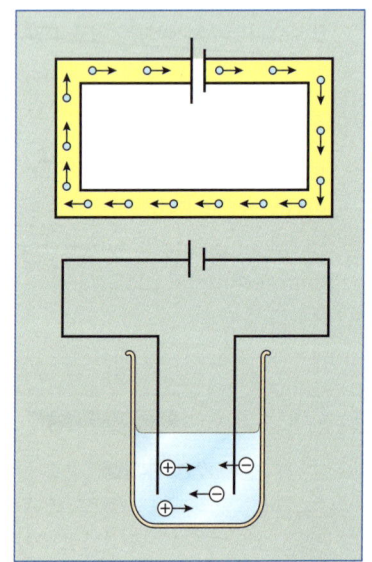

> **charge = current × time or $Q = I \times t$**

In a circuit, the job of the current is to transfer energy from the supply (a battery or the mains electricity) to appliances such as lamps, heaters and motors. Voltage is a measure of the energy transfer for each coulomb of charge:

■ a 12 V battery transfers 12 J of energy to each coulomb of charge

■ if the voltage across a component is 6 V, each coulomb of charge that passes through the component transfers 6 J of energy to it.

◀ 1 volt means "1 joule per coulomb".

The rate of energy transfer is the power supplied to the appliance. High-powered appliances such as heaters need a large current to transfer the energy required each second. The relationship between current, power and voltage is:

> **power = current × voltage or $P = I \times V$**

The total energy transfer by an electric current is equal to the power multiplied by the time:

> **energy transfer = current × voltage × time or $E = I \times V \times t$**

This relationship will be provided on examination pages and module tests.

Motors and transformers

Electric **motors** use energy from the electricity supply to produce movement. They rely on the fact that:

- if there is a current in a wire placed at right angles to the direction of a magnetic field, there is a force on the wire.
- the direction of the force is at right angles to both the current and the magnetic field.
- reversing the current reverses the direction of the force.

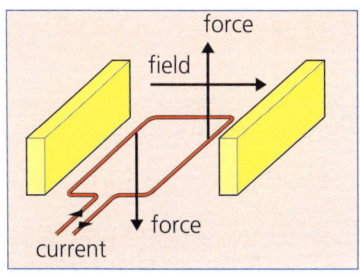
The forces on the wire loop cause it to turn anticlockwise

A **transformer** is a device that changes the size of an alternating voltage. It is made of two coils of wire wound on the same iron core.

The output voltage from the transformer:

- is greater than the input voltage if the secondary coil has more turns than the primary – this is called a step-up transformer;
- is less than the input voltage if the secondary coil has fewer turns than the primary – this is called a step-down transformer.

The relationship between the input and output voltages and the numbers of turns on the coils is:

$$\frac{\text{primary voltage}}{\text{secondary voltage}} = \frac{\text{number of primary turns}}{\text{number of secondary turns}}$$

or $\dfrac{V_\text{p}}{V_\text{s}} = \dfrac{n_\text{p}}{n_\text{s}}$

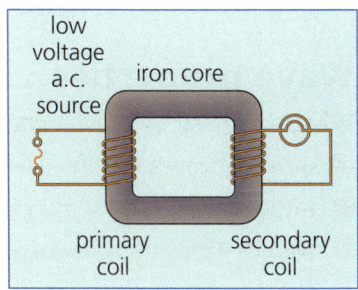
The structure of a transformer

Although you need to be able to recall and use this relationship, you will probably find it easier to use ratios when working out transformer voltages – "the ratio of the voltages is the same as the ratio of the turns".

Delivering power

Electricity is transmitted from power stations to homes and workplaces through a system of cables called the National Grid. Some of these cables are underground and some are overhead, supported by pylons. The cables become heated by the current passing through them. This wastes energy from the electricity supply.

To reduce the amount of energy wasted, the current needs to be as small as possible. This means that the voltage has to be large, so that the required amount of power is transmitted.

The National Grid carries electricity at voltages up to 400 000 V. This would be far too dangerous for householders to use. Transformers are used to:

- increase the voltage and reduce the current at the power station;
- decrease the voltage and increase the current at a substation near to your house.

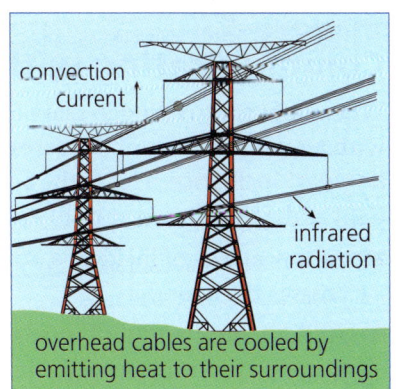
overhead cables are cooled by emitting heat to their surroundings

Questions

1. What is the relationship between moving charge and the size of an electric current?
2. The current in a 240 V mains kettle is 9.5 A. Calculate the power of the kettle.
3. Explain why high voltages are used to transmit electricity.
4. A lamp operates from a 12 V supply. In one minute, 108 C of charge flows through the filament. Calculate the current in the lamp.

Waves and communication

You need to know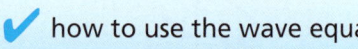

✔ how to use the wave equation;

✔ how optical fibres are used in communication;

✔ how to describe diffraction effects and recall the factors that affect the amount of diffraction that takes place.

Wave properties

Waves are used to transfer energy and information:

- infrared waves transfer energy from a warm object to cooler ones;
- radio waves transfer information to televisions and radios;
- sound waves transfer energy and information to the eardrums.

When waves travel from one place to another, they may cause particles to vibrate, but there is no overall movement of particles between the two places.

The speed of a wave depends on the type of wave it is and what it is travelling through. Electromagnetic waves such as light and radio waves travel much faster than compression waves such as sound.

For all types of wave, the higher the frequency, the shorter the wavelength. The relationship between speed, wavelength and frequency is:

wave speed = frequency × wavelength or $v = f × λ$

> When waves transfer energy from one place to another, there is no overall movement of any material. It is only the vibrations that move.

> You need to know and be able to use this relationship to answer questions in modular tests and written examinations.

Communicating with waves

When light is travelling in glass or water and it meets a boundary with air, whether or not it crosses the boundary depends on the **angle of incidence**. This is the angle between the direction of the light and a line drawn at right angles to the surface.

- At low angles of incidence, some light is reflected and some crosses the boundary.
- At the **critical angle of incidence** the light that escapes forms a spectrum along the boundary.
- At angles of incidence greater than the critical angle, all the light is reflected – this is known as **total internal reflection**.

Total internal reflection is used to bend light round corners in **reflecting prisms** and to contain light within an **optical fibre** so that information can be transmitted over long distances.

Information is usually transmitted along optical fibres using **digital signals**, because these can carry more information than **analogue signals**

> The critical angle for a glass–air boundary is 42°, so light incident at an angle of 45° is totally internally reflected.

All communications are affected by **diffraction**. This describes the way in which waves spread out when they pass through a gap:

■ if the size of the gap is many wavelengths (diagram a), no observable spreading occurs;

■ if the gap is several wavelengths wide (diagram b), there is some spreading into the "shadow" region;

■ if the size of the gap is equal to the wavelength (diagram c), the waves spread so that they can be detected in all directions.

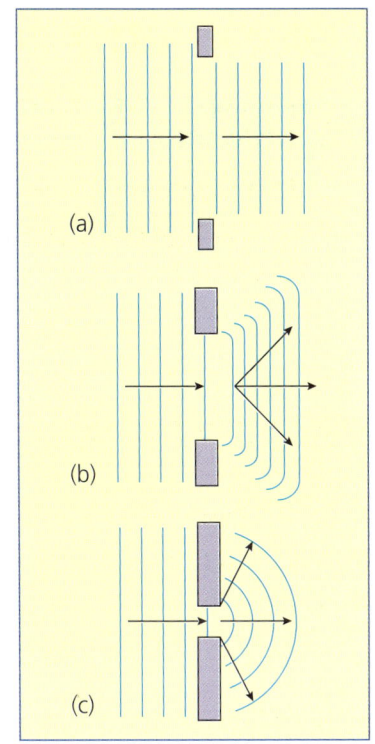

Common effects of diffraction include:

■ Sound spreads out as it passes through a doorway – the size of the gap is comparable to the wavelength.

■ Light does not spread out as it passes through a doorway – the size of the gap is millions of wavelengths of light. Instead, light travels in straight lines and forms shadows.

■ Water waves spread out as they pass through the entrance into a harbour.

Radio and television reception are affected by both diffraction and reflection. Diffraction causes long wavelength radio waves to spread out as they pass over hills and through gaps between buildings so that they can be received in "shadow regions".

The waves that carry television signals and FM radio have a much shorter wavelength. They show very little spreading as they pass round obstacles and through gaps, so reception in hilly areas and cities can be poor.

Poor reception can also be caused by large buildings reflecting radio waves. The radio or television receiver detects the reflection a short time after it receives the direct signal. This causes a fuzzy sound and picture. In extreme cases of television reception it can cause a "ghost" image on the screen.

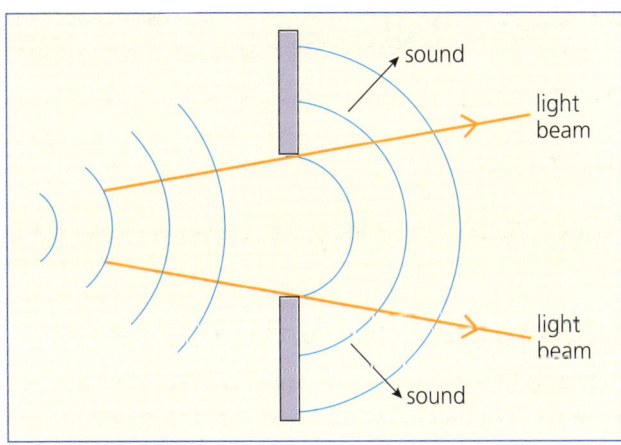

Light travels in straight lines through a doorway but sound spreads out – this shows how diffraction depends on the wavelength of the waves

Questions

1 Calculate the frequency of a wave that has a wavelength of 2.2 m and travels at a speed of 2970 m/s.

2 What is the condition needed for total internal reflection to take place when light meets a glass–air surface?

3 Explain why light travels as a straight beam after passing through a gap that is 5 mm wide but sound spreads out.

4 Name two effects that can reduce the quality of a signal received by a television aerial.

Forces and shape

You need to know ●

✔ how the forces from the supports of a bridge change with the position of the load;

✔ how to describe how the extension of a rubber band or spring depends on the stretching force;

✔ how gas pressure is caused by particle movement.

A simple road bridge consists of a horizontal beam that is supported by two upright concrete pillars. The pillars have to support the weight of the beam as well as any traffic on it. To do this, they exert an upward push on the beam. The greater the load, the greater this push has to be. The two diagrams show the additional forces that act on the bridge because of the weight of the lorry.

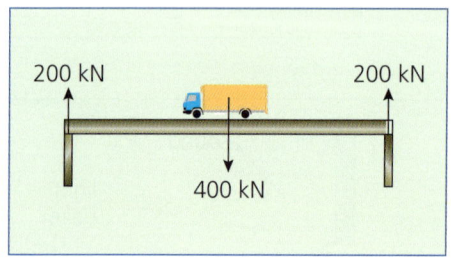

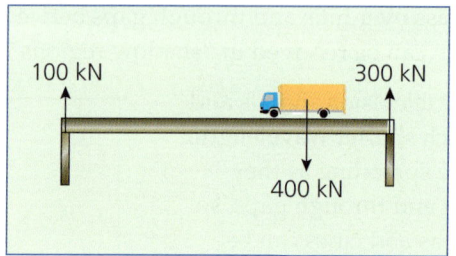

When a load such as a lorry is in the centre of the bridge, the upward forces from the pillars are equal. The nearer the lorry is to one end of the bridge, the greater the upward push from that support.

All materials change shape when they are pulled or pushed. Springs, rubber bands and some wires stretch by a large and measurable amount. The diagrams compare the stretching of these materials.

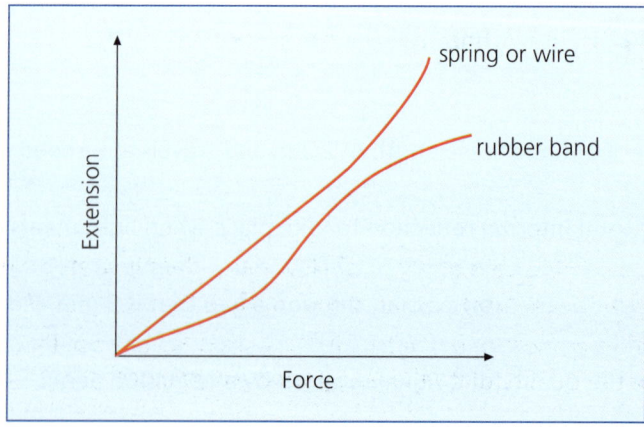

Springs and metal wires behave in a similar way:

- At first the extension is proportional to the force – shown by the straight line part of the graph.
- After this they become easier to stretch but do not return to their original size when the force is removed.

Rubber bands behave differently:

- There is no linear part to the extension–force graph.
- They are difficult to stretch at first, becoming easier to stretch as the force is increased and then more difficult.

Gases consist of large numbers of particles in continual motion. This motion is both rapid and random:

- the particles move at high speeds (on average);
- the speed and direction of any one particle is unpredictable and continually changing.

The motion of gas particles explains how they exert pressure:

- the particles hit the container walls and rebound;
- each collision causes a pushing force on the wall.

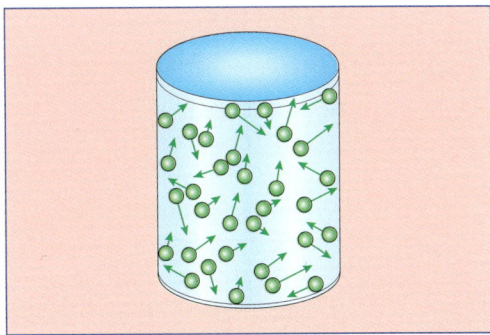

The motion of gas particles is random in both speed and direction

Pressure results from the overall effect of millions of collisions with the container walls each second.

Compressing a gas increases the number of collisions each second and increases the pressure. The relationship between the volume occupied by a gas and the pressure it exerts is:

initial pressure × initial volume = final pressure × final volume or $P_1 \times V_1 = P_2 \times V_2$

Questions

1. A spring stretches by 10 cm when a 20 N force is applied. (a) How much does it stretch when a 5 N force is applied? (b) What have you assumed about the spring?

2. A gas occupies a volume of 50 cm³ at a pressure of 1.00×10^5 Pa. It expands to a volume of 80 cm³ without a change in temperature. Calculate the new pressure of the gas.

Practice module test

1 The current in a lamp filament is 2.0 A when it operates from a 12 V battery. The charge flow through the lamp in one minute is:

- **A** 6 C
- **B** 24 C
- **C** 120 C
- **D** 1440 C

2 Which particles are transferred when a balloon is rubbed with a duster?

- **A** negatively-charged electrons
- **B** positively-charged electrons
- **C** negatively-charged protons
- **D** positively-charged protons

3 In an inkjet printer, charged plates are used to:

- **A** colour the ink drops
- **B** deflect the ink drops
- **C** disperse the ink drops
- **D** reflect the ink drops

4 The current in a toaster element is 5.0 A when it operates from the 240 V mains supply. The power of the element is:

- **A** 48 C
- **B** 48 W
- **C** 1200 C
- **D** 1200 W

5 Which line in the table shows the action of a step-up transformer?

	Input voltage	Output voltage
A	12 V d.c.	240 V d.c.
B	240 V d.c.	12 V d.c.
C	12 V a.c	240 V a.c.
D	240 V a.c	12 V a.c.

6 Electricity is transmitted at high voltage to reduce the energy lost as:

- **A** current
- **B** heat
- **C** power
- **D** voltage

7 Diffraction effects are shown by:

- **A** light but not sound
- **B** sound but not light
- **C** neither light nor sound
- **D** both light and sound

8 Which diagram shows waves being diffracted?

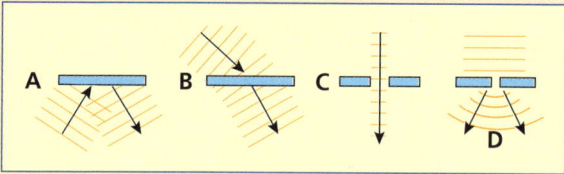

9 A "ghost" image on a television screen is caused by:

- **A** diffraction
- **B** interference
- **C** reflection
- **D** refraction

10 Total internal reflection occurs when:

- **A** the angle of incidence is less than the critical angle
- **B** the angle of incidence is greater than the critical angle
- **C** the angle of reflection is less than the critical angle
- **D** the angle of reflection is greater than the critical angle

11 Gas pressure is a result of:

- **A** the particles being squashed together
- **B** the particles being widespread
- **C** the particles colliding with each other
- **D** the particles colliding with the walls of the container

12 As the lorry in the diagram travels from right to left over the bridge:

- **A** forces P and Q both increase
- **B** forces P and Q both decrease
- **C** force P increases and force Q decreases
- **D** force P decreases and force Q increases

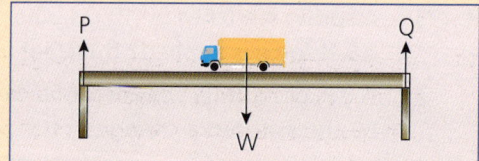

13 A positively charged object is connected to the ground. It discharges by:

  A electrons moving from the object to the ground

  B electrons moving from the ground to the object

  C protons moving from the object to the ground

  D protons moving from the ground to the object

14 Which line in the table shows the sign of the charge carried in a conducting metal and a dissolved electrolyte?

	metal	dissolved electrolyte
A	positive	negative
B	positive and negative	positive
C	negative	positive
D	negative	positive and negative

15 The current in the filament of a 60 W lamp when it operates from the 240 V supply is:

  A 0.25 A

  B 4.0 A

  C 0.25 W

  D 4.0 W

Questions 16 and 17

9000 C of charge flow through a 12 V battery in 5 minutes.

16 The current in the battery is:

  A 2.5 A

  B 5.0 A

  C 30 A

  D 150 A

17 The energy transferred by the charge is:

  A 360 J

  B 750 J

  C 108 kJ

  D 540 kJ

18 A transformer is used to step up the 240 V mains voltage to 48 kV. Which line in the table gives possible values of the numbers of turns on the coils?

	Turns on primary coil	Turns on secondary coil
A	20	400
B	20	4 000
C	400	20
D	4 000	20

19 The speed of radio waves is 3.0×10^8 m/s. A radio station broadcasts on a wavelength of 1500 m. The frequency of the waves is:

  A 200 Hz

  B 200 kHz

  C 200 MHz

  D 200 s

20 A digital signal:

  A has a higher frequency than an analogue signal

  B can carry more information than an analogue signal

  C has a longer wavelength than an analogue signal

  D travels faster than an analogue signal

21 Light of wavelength 5.0×10^{-7} m can be diffracted by passing it through a slit of width:

  A 2.5×10^{-6} m

  B 2.5×10^{-4} m

  C 2.5×10^{-2} m

  D 2.5 m

22 Diffraction of radio waves:

  A prevents reception in hilly areas

  B can enable reception in hilly areas

  C prevents reception in flat areas

  D can enable reception in flat areas

23 The motion of particles in a gas:

  A is random in speed but orderly in direction

  B is orderly in speed but random in direction

  C is orderly in both speed and direction

  D is random in both speed and direction

24 When the volume of a gas is doubled at a constant temperature, its pressure:

  A quarters

  B halves

  C doubles

  D quadruples

Answers to these questions can be found on pages 143–147

Getting it right

(a) The diagram shows a balloon and a duster, after the balloon has been rubbed with the duster.

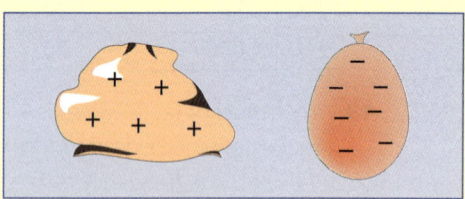

(i) Explain how the balloon and the duster become charged

Electrons move from the duster to the balloon. **[2]**

(b) When an aircraft is being refuelled, charge can build up on the metal body of the aircraft.

(i) Explain how the whole of the body of the aircraft becomes charged.

The metal body conducts electricity. Charge can move through a conductor. **[2]**

(ii) Suggest why this build-up of charge could be dangerous. Give reasons for your answer.

The charge could create a high voltage. This could cause a spark, which could ignite the fuel. **[3]**

(iii) Explain how earthing the body of the aircraft reduces this danger.

Earthing allows the aircraft to discharge by a flow of electrons to the object at a higher positive voltage. **[3]**

The extra mark here is given for the quality of written communication. The answer is clear and logical, with correct punctuation and grammar. You can recognise where marks are awarded for the quality of written communication by the icon in the margin.

Exam questions

Question 1B

The flow diagram below shows the cells involved in human reproduction, and the numbers of chromosomes in the body cells of Mr and Mrs Lewis.

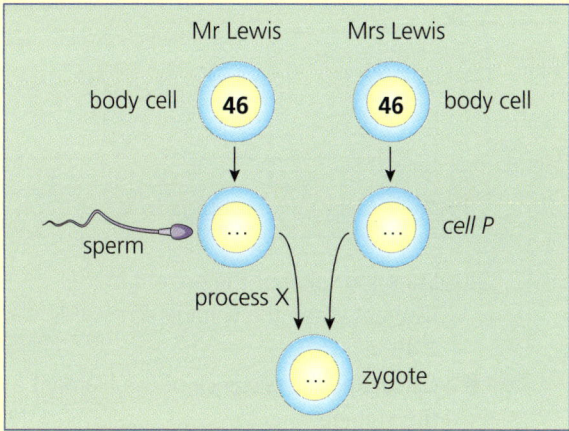

a What are the chromosome numbers in the sperm, cell P and the zygote? [3]

b The diagram shows the sperm fusing with cell P.
   i Name cell P. [1]
   ii Name process X. [1]

c The parental cells each contain 23 pairs of chromosomes. What term describes this chromosome number? [1]

d The sex chromosomes for Mr and Mrs Lewis are shown. Explain how Mr and Mrs Lewis can produce a boy **or** girl

   Mr Lewis   ×   Mrs Lewis
      XY   XX   [3]

Question 1C

This question is about the Earth's atmosphere. The composition of the atmosphere is kept constant because some processes use up oxygen (respiration and combustion) and others produce oxygen (photosynthesis).

a What is the name of the gas present in the Earth's atmosphere today in the largest amounts? [1]

b According to a recent magazine article, 300 years ago the atmosphere contained 38% oxygen. Today it contains about 20%. Explain why this change may have happened. [4]

c How has the amount of carbon dioxide in the atmosphere changed over the same period?. [2]

Question 1P

The diagram shows the horizontal forces acting on a car.

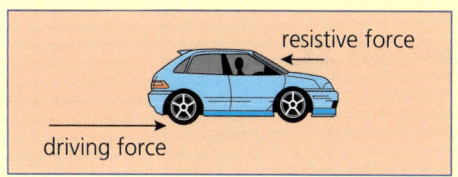

a Explain how you can tell from the diagram that the car is accelerating in a forward direction. [2]

b The car increases its speed from 5 m/s to 15 m/s in 4.0 s. Calculate the acceleration of the car. [3]

c How does the increased speed of the car affect:
   i The driver's reaction time? [1]
   ii The thinking distance when the driver sees a hazard? [1]
   ii The braking distance when the driver applies the brakes? [1]

Question 2B

The diagram shows structures involved in sweating.

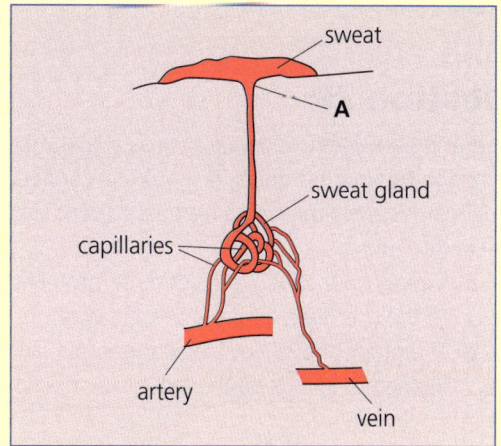

a Name part A. [1]

b Sweat is shown on the surface of the skin.
   i Name two substances contained in sweat. [2]

ii How do substances in sweat reach the
   sweat gland? [1]
iii Explain how the sweat cools the body
    down. [3]

c Antiperspirant sprays stop you sweating from
  the areas of skin that they cover. What effect
  could this have on body temperature? [1]

d Explain how the skin prevents entry of
  harmful microorganisms into the body. [3]

Question 2C

Poly(chloroethene) is a polymer used to make the
insulation on electrical cables.

a Suggest two properties of poly(chloroethene)
  that make it suitable for this use. [2]

b Suggest one other use of poly(chloroethene). [1]

c The monomer used to make
  poly(chloroethene) is called chloroethene. It
  has a molecular formula of C_2H_3Cl.

  i Describe how you would show that
    chloroethene contains a carbon–carbon
    double bond. [2]

  ii Draw the structures of chloroethene and
     ethane. [2]

  iii Draw the structure of poly(chloroethene). [2]

d Burning poly(chloroethene) can produce
  strongly acidic gases. These are not formed
  when poly(ethene) is burned. Suggest the
  name of the chemical in these fumes. [1]

Question 2P

When using an electrical device such as a lawnmower
or a hedge trimmer outdoors, it should be connected
to the mains supply through a residual current circuit
breaker (RCCB).

a Explain how an RCCB gives more protection
  to the user than a fuse and earth wire. [3]

b The graph shows how the current in a lamp
  filament depends on the voltage across it.

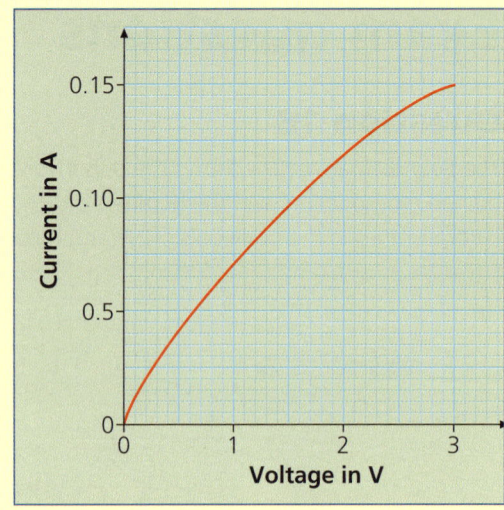

i State the value of the current in the
  filament when the voltage across it is
  2.40 V. [1]

ii Calculate the resistance of the filament at
   this voltage. [3]

iii What does the graph show about how
    the resistance changes as the voltage is
    increased? Explain how you can tell this
    from the graph. [2]

Question 3B

a The following statements list the stages in
  the transfer of a human gene to a bacterium,
  but they are in the wrong order.
  A The section of human DNA is inserted into
    the bacterial plasmid
  B The plasmid with the human DNA is
    cloned inside the bacterium
  C A section of human DNA is removed
  D The plasmid is removed from a bacterium
  E Human DNA is cut with restriction enzyme
  F The plasmid DNA is also cut with
    restriction enzyme
  What is the correct order for this example of
  genetic transfer? [6]

b Scientists transferred a gene for resistance to
  weed-killer from wheat to soya bean plants.
  What is the advantage to farmers growing
  this GM crop? [4]

Answers to these questions can be found on page 148

Answers to end of spread questions

Module 1

END OF SPREAD QUESTIONS

P3

1. Suitable temperature and pH.
2. It is denatured or destroyed (NOT killed!)
3. The reaction is slower and enzyme eventually becomes inactive.
4. Nothing happens to protein in the mouth – there is no protease enzyme present.
5. The small intestine is very long, so has a large surface area. There are large numbers of villi and each has many microvilli.

P5

1. (a) Phagocytes engulf microorganisms and ingest them. They react to antigens (foreign proteins). (b) Lymphocytes also react to antigens, and produce antibodies.
2. Red cells have a lot of haemoglobin, which has an affinity for oxygen. They produce oxyhaemoglobin. They have a high surface area and the round shape helps them to move through the blood vessels.
3. Lens – short and fat. Suspensory ligaments – loose. Ciliary body – contracted.
4. Too much light entering the eye can damage the retina. Too little light and you cannot see.
5. A sedative. Cirrhosis (hardening of the liver) or brain damage.

P7

1. Reabsorption is when substances like glucose go from the kidney nephron back into the blood.
2. Through the first convoluted tubule a lot of water is reabsorbed, so the filtrate becomes concentrated.
3. To maintain an internal balance. At the wrong temperature or pH reactions would not take place. Too much or too little of a chemical produced would have a bad effect on the body.
4. Bowman's capsule, first convoluted tubule, (loop of Henle), second convoluted tubule, collecting duct, ureter, bladder, urethra.

P9

1. Lysozyme destroys bacteria.
2. Sweat excreted onto the skin surface, water from sweat evaporates, this needs body heat, so we cool down.
3. Vasodilation.

PRACTICE MODULE TEST

1. A	5. C	9. B	13. A
2. B	6. C	10. C	14. C
3. B	7. A	11. C	15. C
4. D	8. B	12. D	16. D

Module 2

END OF SPREAD QUESTIONS

P15

1. X and Y. XX = female, XY = male.
2. (a) 46. (b) 23.
3. Meiosis.
4. 2, 4.

P17

1. (a) Guanine, (b) adenine.
2. To map the human genomes along the chromosomes/map human DNA.
3. The production of new individuals without using any sex cells.

P19

1. Thick waxy cuticle to reduce water loss; spines to avoid being eaten by herbivores; wide network of roots to absorb the little water available; thick stem to store water.
2. (i) Not enough plants to eat.
   (ii) Eaten by predators.
   (iii) Killed by disease.
3. There may be a change in the environment. As a result some organisms are killed. Others survive and go on to pass on advantageous genes to offspring.
4. Plants compete for light, water and minerals.
5. A change in a gene or chromosome or a change in DNA.

P21

1. The production of substances and/or processes that are harmful to the environment.
2. Acid rain is caused by the burning of fossil fuels. This produces sulphur dioxide, which dissolves into moisture in clouds, forming sulphurous acid.
3. Human population increase. More people need more goods/produces so pollution increases.

1.	C	6.	B	11.	A	15.	A
2.	C	7.	D	12.	B	16.	D
3.	C	8.	B	13.	B	17.	C
4.	B	9.	B	14.	B	18.	C
5.	A	10.	C				

Module 3

END OF SPREAD QUESTIONS

P27

1. Protons and neutrons
2. Electron
3. Electron
4. Phosphorus
5. Non metal
6. Potassium
7. 6

P29

1. Bromine
2. Potassium chloride + bromine
3. Displacement reaction
4. Chlorine
5. $Cl_2 + 2KBr \rightarrow 2KCl + Br_2$
   $Cl_2 + 2Br^- \rightarrow 2Cl^- + Br_2$

P32

1. Catalyst
2. Shorter time
3. Enzyme
4. Faster

PRACTICE MODULE TEST

1.	C	6.	C	11.	B	16.	C
2.	D	7.	C	12.	D	17.	A
3.	C	8.	B	13.	A	18.	A
4.	A	9.	A	14.	C	19.	B
5.	C	10.	D	15.	C	20.	D

Module 4

END OF SPREAD QUESTIONS

P39

1. Carbon and hydrogen
2. Fractional distillation
3. Higher temperature, high pressure, lack of oxygen
4. Smaller the molecules, the higher in the tower

P40

1. Carbon dioxide and water
2. Carbon monoxide
3. $CH_4 + O_2 \rightarrow C + 2H_2O$

P43

1. Alkane
2. 2n + 2
3. Cylcohexene turns bromine colourless. No change with cyclohexane.
4. Products are more valuable

P44

1. $C_6H_{12}O_6 \rightarrow 2C_2H_5OH + 2CO_2$
2. Enzymes denatured at 70 °C

P45

1. Glass, cement and iron
2. Sulphuric acid
3. Fertiliser
4. Oxidation

PRACTICE MODULE TEST

1.	C	6.	B	11.	D	16.	A
2.	B	7.	C	12.	B	17.	C
3.	A	8.	A	13.	C	18.	B
4.	D	9.	B	14.	A	19.	B
5.	C	10.	B	15.	B	20.	A

Module 5

END OF SPREAD QUESTIONS

P51

1. Direct current passes in the same direction all the time. Alternating current changes direction.
2. In parallel.
3. $4.8\,\Omega$.
4. LDR.

P53

1. The live conductor.
2. The earth.
3. A difference in the currents carried by the live and neutral wires.

P55

1. They rotate inside copper coils.
2. Transformers.
3. Fossil fuels.
4. It is a poor conductor and cannot form convection currents.

PRACTICE MODULE TEST

1.	B	7.	D	13.	C	19.	B
2.	C	8.	A	14.	A	20.	D
3.	B	9.	B	15.	C	21.	D
4.	D	10.	B	16.	B	22.	B
5.	C	11.	D	17.	A	23.	C
6.	D	12.	A	18.	C	24.	D

Module 6

END OF SPREAD QUESTIONS

P61

1. 630 m
2. Ultrasound has a higher frequency or cannot be heard by humans.
3. Gamma rays are high-energy radiation that can kill microbes.
4. Unlike X-rays, ultrasound does not damage body tissue.

P63

1. Jupiter is very massive.
2. The force increases in size as the comet approaches the Sun. The force changes direction so that it is always directed towards the Sun.
3. The galaxy is moving towards the Earth.

P65

1. They have the same number of protons and different numbers of neutrons.
2. The alpha radiation cannot penetrate the casing of the smoke alarm.
3. The decay of any individual nucleus is unpredictable.

PRACTICE MODULE TEST

1. D	7. C	13. A	19. B
2. A	8. D	14. B	20. D
3. A	9. C	15. A	21. B
4. C	10. B	16. C	22. D
5. D	11. A	17. A	23. D
6. B	12. C	18. C	24. B

Module 7

END OF SPREAD QUESTIONS

P71

1. **a** xylem **b** phloem
2. Movement of water molecules through a selectively permeable membrane from a lower concentrated solution to a higher concentrated solution.
3. Guard cells
4. Used to release energy, used in respiration, changed to starch, changed to cellulose, changed to fats or oils, changed to proteins.

P73

1. Producers
2. Energy lost by respiration and excretion (or named form of excretion).

3. Moloculture is growing one crop only, year after year in the same field. This reduces biodiversity because the animals have such a limited food type.

P75

1. Warmth and wet
2. To prevent build up of waste/ to recycle/ to return useful minerals to soil.
3. Decomposers use aerobic respiration. Giving them air enables the oxygen to ensure decay rate at a maximum.
4. Respiration, photosynthesis, combustion.
5. Clearing ground for agricultural crops, or housing or factories.
6. Certain bacteria are able to use atmospheric nitrogen to make ammonium compounds. These bacteria are found with root nodules of plants of the pea and bean family. The plants are able to make proteins using the ammonium compounds.

P77

1. Weeds killed so there is more light, more minerals, more water fro the crops.
2. No harmful chemicals are found in the food products.
3. Biological control often used a predator. The pesticide may kill the predator and the pest so that biological control does not work.
4. Longer growing period only to optimum temperature conditions.

PRACTICE MODULE TEST

1. B	6. A	11. D	16. A
2. A	7. A	12. D	17. B
3. B	8. C	13. C	18. C
4. C	9. C	14. C	19. A
5. B	10. A	15. D	

Module 8

END OF SPREAD QUESTIONS

P83

1. Diaphragm contracts and flattens.
2. Alveoli are a very large surface area which enable oxygen to diffuse into blood and carbon dioxide to diffuse out.
3. Air in nose is filtered so less dust and microbes reach lungs. Air through nose is warmed. Less of a temperature shock for the lungs.

P85

1. Insulin reduces blood glucose. Glucose is allowed into cells to be used to release energy. Excess glucose enters liver to be made into glycogen.
2. Polyunsaturated fats are lower in cholesterol so atheroma (blockage of vessels by fatty deposits) less likely.
3. Where heart is deprived of oxygen - gives a sharp pain.

P87

1. Glucose and oxygen
2. Rate of aerobic respiration will increase, because more oxygen supplied to cells.
3. Lactic acid.

P89

1. Cystic fibrosis
2. Let N = cystic fibrosis trait
   n = normal

	N	n
N	NN	Nn
n	Nn	nn

   There is a 1 in 4 chance of producing a cystic child.
3. Gene therapy is inserting genes into cells. This is often done to overcome symptoms. In cystic fibrosis an aerosol is used in the lungs.

PRACTICE MODULE TEST

1.	B	6.	A	10.	B	14.	D
2.	A	7.	B	11.	B	15.	B
3.	C	8.	C	12.	B	16.	A
4.	C	9.	B	13.	C	17.	B
5.	D						

Module 9

END OF SPREAD QUESTIONS

P95

1. Sodium
2. iron ore, coke, limestone
3. reduction

P96

1. C
2. A
3. $Fe^{2+} \rightarrow Fe^{3+} + e^-$ oxidation

P97

1. Sodium hydroxide
2. $2Li + 2H_2O \rightarrow 2LiOH + H_2$
   $2K + 2H_2O \rightarrow 2KOH + H_2$
3. More reactive

P99

1. Limestone
2. Marble
3. Granite
4. Y

P101

1. Nitrogen
2. Oxygen
3. Reduces photosynthesis – less oxygen in the air/more carbon dioxide

PRACTICE MODULE TEST

1.	A	6.	B	11.	A	16.	C
2.	D	7.	A	12.	A	17.	B
3.	D	8.	B	13.	A	18.	D
4.	C	9.	B	14.	D	19.	B
5.	A	10.	D	15.	D	20.	C

Module 10

END OF SPREAD QUESTIONS

P107

1. 15
2. 15
3. 16
4. 8
5. Oxygen – 16 8   Oxygen – 18 10
6. 2
7. 152

P110

1. **a** ionic
   b covalent
2. Ionic
3. Covalent
4. True
5. Giant structure

P113

1. Glucose + oxygen $\rightarrow$ carbon dioxide + water + energy
2. Endothermic
3. No
4. 12
5. 2
6. 44
7. 17
8. 3.2g
9. 8.0g
10. CH_4

PRACTICE MODULE TEST

1. C	6. B	11. B	16. A
2. C	7. D	12. C	17. C
3. C	8. A	13. D	18. D
4. D	9. A	14. D	19. B
5. B	10. B	15. A	20. A

Module 11

END OF SPREAD QUESTIONS

P120

1. Terminal velocity.
2. 1.25 m/s^2.
3. The reaction time is the same, so the distance travelled is proportional to the speed.

P121

1. 4500 J.
2. Kinetic energy to gravitational potential energy.
3. 83 300 J.

P123

1. Continental plates moving away from each other.
2. P-waves or longitudinal waves.
3. S-waves cannot travel through the liquid outer core.

P125

1. **a** 600 Bq.
   b 300 Bq.
2. Cloth and wood.
3. 30 s.

PRACTICE MODULE TEST

1. D	7. D	13. D	19. D
2. C	8. A	14. D	20. A
3. B	9. C	15. C	21. A
4. D	10. D	16. D	22. D
5. B	11. C	17. C	23. B
6. D	12. A	18. C	24. A

Module 12

END OF SPREAD QUESTIONS

P131

1. **a** negative
   b positive
2. It provides a conducting path for charge to pass to the ground.
3. Charge can spread out, preventing the build up of a high voltage.

P133

1. The size of a current in A is equal to the rate of flow of charge in C/s.
2. 2280 W
3. To minimise the current, reducing the power losses.
4. 1.5 A.

P135

1. 1350 Hz.
2. The angle of incidence is greater than the critical angle.
3. The gap is many wavelengths wide for light but less than a wavelength for sound.
4. Reflection and diffraction.

P137

1. (a) 2.5 cm. (b) That a force of 20 N does not stretch the spring beyond the linear part of the extension–force graph.
2. 6.25×10^4 Pa.

PRACTICE MODULE TEST

1. C	7. D	13. B	19. B
2. A	8. D	14. D	20. B
3. B	9. C	15. A	21. A
4. D	10. B	16. C	22. B
5. C	11. D	17. C	23. D
6. B	12. C	18. B	24. B

Answers to exam questions

Question 1B

a *Sperm = 23; cell P = 23; zygote = 46.*

b i *Ovum/egg cell.*
   ii *Fertilisation.*

c *Diploid.*

d
                    sperms

ova		X	Y
	X	XX	XY
	X	XX	XY

  50% boys, 50% girls.

Question 1C

a *Nitrogen.*

b *Increased population; more respiration/more carbon dioxide produced; much greater use of (fossil) fuels/coal; uses up oxygen and produces carbon dioxide; increased pollution; kills algae/plants so reducing photosynthesis/less oxygen released/destruction of forests/more buildings (one mark each for any three points, plus one mark for answer written in sentences with correct punctuation and grammar).*

c *More carbon dioxide; decreasing oxygen is linked to increasing carbon dioxide.*

Question 1P

a *The driving force is greater than the resistive force; so there is an unbalanced force in the forward direction.*

b *Acceleration = increase in velocity ÷ time taken (or symbols)*
$$= 10 \text{ m/s} \div 4.0 \text{ s}$$
$$= 2.5 \text{ m/s}^2$$
  (one mark for each line)

c i *No effect.*
   ii *Increases.*
   iii *Increases.*

Question 2B

a *Sweat pore.*

b i *Urea, water, sweat (any two).*
   ii *Through the blood/through capillaries.*
   iii *Sweat lies on skin; evaporation takes place; heat energy from body needed (for evaporation to take place); so body cools down (any three).*

c *Body temperature increases.*

d *Skin cells are tough/have keratin; form a barrier; sebum/oil produced; contains a chemical that kills bacteria/contains lysozyme (any three).*

Question 2C

a *Non-conductor of electricity; easily moulded; easily coloured; does not rot (any two).*

b *Gutters and drainpipes; clothes; Wellington boots (any one).*

c i *Add bromine (solution). Solution turns from brown to colourless/is decolourised.*
   ii

  chloroethene   ethane

  (one mark each)
   iii

  (one mark for structure showing a chain and one for only a single bond between carbon atoms)

d *Hydrogen chloride or hydrochloric acid.*

Question 2P

a *An RCCB detects any difference in the currents in the live and neutral; it breaks the circuit if there is a tiny leak to earth; a fuse acts much more slowly; it would not blow until there was a large current to earth (any three).*

b i *0.135 A.*
   ii *Resistance = voltage ÷ current (or symbols)*
$$= 2.40 \text{ V} \div 0.135 \text{ A}$$
$$= 17.8 \ \Omega$$
  (one mark for each line)
   ii *The resistance increases; as the voltage increases the current also increases but at a decreasing rate.*

Question 3B

a *E, C, D, F, A, B.*

b *They can use weed-killer on soya bean fields; weeds compete with the crop; for light/for water/for minerals; so yield maximum.*

Getting the grade you deserve

Exams are an opportunity for you to show just how much you have learnt in all your science lessons. It would be a great shame if you didn't get all the marks you deserve.

Knowing what to expect in the exam will help you make the most of your scientific skills and knowledge. The advice on these pages will help you to understand what the examiner is asking you to do and help you to prepare in the best way possible.

What types of questions will I have to answer?

You will find a number of different types of questions in each exam paper. These include:

- sentence completion – you may or may not be given the responses to choose from;
- short-answer questions – these can be answered with a single word or phrase or by linking boxes;
- data handling questions – these give you information, usually in the form of a table or a graph, that you have to use to answer the question;
- long-answer questions – these need you to make at least two points in one or more sentences and are worth at least two marks;
- calculations – you need to recall or select the correct relationship and use it correctly to calculate the size of a quantity.

How should I start to answer a question?

Many questions begin with a short piece of introductory material that gives you some information about a scientific context and may include a diagram. This introduction often contains key information that you need to answer the question.

Many candidates do not bother to read the introduction to a question. They go straight to the questions being asked and then find that they do not know the answers. The answers may be staring you in the face!

- Read the introduction carefully. If it's lengthy read it two or three times.
- Underline key points.
- Examiners go to great lengths to put the least number of words possible in a question. They only put things in that are important.
- When you have put together an answer in your head, refer back to the introductory material before putting pen to paper.

How can I tell what I have to do to gain the most marks? • • • • • • • • • • • • • • • •

First, always look at how many marks are allocated to the answer. This is printed in brackets after the answer space.

■ If there are two marks then you need to make *two* separate correct points to gain them both.

■ If there is only one mark and you give two answers, one of which is correct and one of which is wrong, you will not be awarded any marks for your answer.

Then, look at the cue word that the question begins with. The most common cue words are: **State** or **give** or **name** A single word, phrase or sentence is needed.

Describe

This requires a description of:

■ a scientific process;

■ how one variable is related to another;

■ a trend or pattern in data – this will be given either in a table or on a graph

Explain

This is asking you to give reasons to support your answer. A simple description is likely to gain only one mark out of a possible three. Your answer should include "because" followed by the reasons. Check carefully to see if you really have "explained".

Suggest

Suggest can be asking for a description but **suggest why** is usually asking for an explanation. Suggest is used instead of describe or explain for a number of reasons:

■ the question might involve an application of science that you are not required to be familiar with;

■ there might be no clear correct answer;

■ the question might ask you to make a prediction.

Whatever the case, you are expected to use your knowledge and understanding of science to answer the question.

How should I approach calculations? • • •

Calculations should always be set out in a clear, logical order. There may be a "quality of written communication" mark just for doing this. The steps to take are:

■ Write down the relationship that you are using in either words or standard symbols, whether it is one you need to recall or one that is given on the exam paper.

- Write out the relationship with the correct quantities written in the place of the words or symbols.
- Calculate the final answer and state the correct units.

One mark out of the 30 on each exam paper requires knowledge of a correct scientific unit.

How are the marks awarded for drawing a graph?

If you are given a blank grid to draw a graph on, four marks are available:

- one mark for choosing appropriate scales and labelling the axes with the quantity and unit – for example "voltage in volts";
- two marks for plotting the points correctly;
- one mark for drawing the best-fit straight line or smooth curve.

Points to note when drawing graphs are:

- use a pencil so that you can erase mistakes;
- make each plotted point clearly visible – dots are easily hidden behind a line, so use crosses;
- never draw a line by joining the plotted points together, use a smooth curve or line of best fit;
- never draw a line by joining the first and last points together – there is no reason why these should be any more reliable than the others;
- use a transparent ruler to draw a straight line, and judge it so that there are as many points above the line as there are below it;
- check any points that seem to be a long way from the line – you may have made a mistake.

Assessing ideas and evidence

What's it worth?

The assessment of ideas and evidence in science has to be worth 5% of all the available marks. However, coursework and module tests do not examine this, so all of the assessment is done on the final exam papers. This means that 10% of the marks on each paper, or three marks out of 30, are devoted to assessing this aspect of your course.

What are the examiners looking for?

There are four aspects of ideas and evidence in science that will be examined:

1. How scientific ideas are presented, evaluated and disseminated.

    This is straightforward. Questions will ask you how scientists communicate with other scientists and make their ideas known throughout the world. The answers are through international conferences, and writing in books and magazines such as *Nature*, *Scientific American* and *New Scientist*.

2. How scientific controversies can arise from different ways of interpreting empirical evidence.

    The phrase "empirical evidence" means evidence gathered from experiments or data collection. Questions that assess this aspect of ideas and evidence are likely to be about situations where the scientific evidence is unclear – for example the evidence linking electricity power lines or nuclear power stations to cancer. You should be prepared to give balanced arguments, emphasising the different ways in which the evidence can be judged.

3. Ways in which scientific work may be affected by the context in which it takes place

    A good example of this is the cloning of animals and the possibility of cloning humans. You will not be expected to know all the details of "Dolly the sheep", as these will be given to you in the question. You will be expected to give the arguments about whether human cloning should be allowed, or whether human fetuses should be developed just to obtain important biochemicals from them.

4. Ways to consider the power and limitations of science in addressing industrial, social and environmental questions, including the kinds of questions science can and cannot answer, uncertainties in scientific knowledge and the ethical issues involved.

    There are lots of topical issues and examples in recent history of questions that science cannot answer or has got the answers wrong. Examples include the BSE and foot and mouth crises.

You need to be able to give reasons as to why science can or cannot provide the answers to these. The reasons may be that scientists are often unable to experiment on humans or do not have enough evidence to predict the effects of action or inaction.

How will I know which questions assess ideas and evidence?

Unlike the assessment of written communication, there will not be an icon on exam papers to show where ideas and evidence is being assessed. Sometimes it will be clear, for example "How do scientists communicate their findings to other scientists?" At other times it will not be so clear.

The questions assessing ideas and evidence will always be embodied within other questions. Remember, keep an open mind, be prepared to make suggestions and put together a logical argument from both points of view when answering questions about issues where there is no definitive answer.

Index